W0227761

Konrad Maurer · Thomas Dierks

Atlas of
Brain Mapping

Topographic Mapping of EEG
and Evoked Potentials

With 66 Figures

Springer-Verlag
Berlin Heidelberg New York
London Paris Tokyo
Hong Kong Barcelona
Budapest

Prof. Dr. Konrad Maurer
Dr. Thomas Dierks

Psychiatrische Univ. Klinik Würzburg
Füchsleinstr. 15
W-8700 Würzburg, FRG

ISBN-13: 978-3-642-76045-7   e-ISBN-13: 978-3-642-76043-3
DOI: 10.1007/978-3-642-76043-3

This work is subject to copyright. All rights are reserved, whether the whole or part of the material is concerned, specifically the rights of translation, reprinting, reuse of illustrations, recitation, broadcasting, reproduction on microfilm or in other ways, and storage in data banks. Duplication of this publication or parts thereof is only permitted under the provisions of the German Copyright Law of September 9, 1965, in its current version and a copyright fee must always be paid. Violations fall under the prosecution act of the German Copyright Law.

© Springer-Verlag Berlin Heidelberg 1991
Softcover reprint of the hardcover 1st edition 1991

The use of general descriptive names, registered names, trademarks, etc. in this publication does not imply, even in the absence of a specific statement, that such names are exempt from the relevant protective laws and regulations and therefore free for general use.

Product liability: The publishers can give no guarantee for information about drug dosage and application thereof contained in this book. In every individual case the respective user must check its accuracy by consulting other pharmaceutical literature.

Typessetting, printing and binding: Appl, Wemding
25/3130-543210 – Printed on acid-free paper

Dedicated to our patients, for whom this method
should provide improved diagnosis and possibilities for treatment

Preface

From its discovery in 1929 by Hans Berger until the late 1960s, when sensory visual and auditory evoked potentials were discovered and became popular, the EEG was the most important method of neurophysiological examination. With the advent of computer technology in the 1980s, it became possible to plot the potential fields of the EEG onto models of the scalp. This plotting of information as neuroimages followed the structural and functional techniques of CT, MRI, PET and SPECT. The success of this method, which began in the early 1980s, has led to the brain mapping of EEGs and EPs being increasingly used for diagnosistic purposes in neurology, psychiatry and psychopharmacology.

The pioneers of this method believed in it and were committed to its success. However, many traditionalists felt that it gave no new information and so regarded the method with scepticism. Some found both the coloured maps and the mapping technique misleading, which led to unnecessary conflict between mappers and their chromophobic oponents. Emotions have run so high that some professional bodies have justifiably adopted guidelines and warned of the misuse of the method.

As mapping is still in the process of change, it is one of our aims to describe the techniques, clinical applications and the results of mapping in an easily understandable way, so that there can be informed discussion concerning the theory and techniques which are used in mapping. It is recognised that the techniques involved in mapping, which are based on quantitative EEG analysis and computer technology, can only be successfully applied if there is a sound understanding of the EEG and evoked potentials. This atlas explains brain mapping techniques clearly and shows how maps can be used to illustrate complex and difficult problems. An advantage of the best mapping systems is that they simultaneously record both the EEG and evoked potentials.

A further advantage of EEG mapping is that it uses computer technology to quantify the EEG and plot out the results of this analysis in understandable form. It also employs statistical tests to give significance to the analysed data. This is more rigorous than conventional EEG and evoked potential analysis, which does not demand such a comprehensive understanding of the electrical waveforms and their display. The raw data which goes into mapping must be of the highest quality, and this places extensive demands on the technicians.

To quote from an entry Berger made in his diary in Jena on November 16th 1924, "May I succeed in achieving my plan of more than 20 years and create a kind of brain mirror, the electroencephalogram!". Initially the EEG did not fulfil the role of a 'brian mirror' because of the limitations of technology and our understanding of brain function. These limitations were highlighted by the apparent success in the diagnosis of structural pathology made by CT and MRI and functional changes as measured by PET. It has yet to be shown whether mapping techniques will be able to fulfil Berger's expectations of the EEG and provide a true 'brain mirror'. However, with the advances in our understanding of brain mapping it is at last conceivable that Berger's dream may become reality.

We could never have compiled this atlas without help from many different people. We would especially like to thank I. Gröbner and S. Gahn who so accurately made the recordings of both patients and controls, Mrs. Möslein who prepared and corrected the manuscript and Professor Morice from the University of Newcastle, Australia, who read the text with great care.

We would also like to thank our families for their care and patience and for having managed without us.

Springer-Verlag showed its usual expertise in the production of this atlas, particularly its magnificent color illustrations. Our special thanks goes to Dr. T. Thiekötter, who showed great interest in the concept of this atlas and whose support helped us to complete it. We would also like to thank the staff of Springer-Verlag, especially S. Benko, B. Löffler, and Dr. M. Wilson, who provided valuable help in the publication of this book. We are also grateful to Schwind Medizin-Technik, Erlangen, whose generous support made it possible to produce an atlas of this high quality.

July 1991   KONRAD MAURER
                                                             THOMAS DIERKS

Contents

1 Introduction 1

2 History 3

3 Definition and Terminology 7

4 Methodology 9
   4.1   Introduction 9
   4.2   General Conditions 9
   4.3   Calibration 11
   4.4   Electrodes 11
   4.5   References 14
   4.6   Baseline 18
   4.7   Artifacts 18

5 Data Acquisition and Signal Analysis 23
   5.1   Analog to Digital Conversion 23
   5.2   Aliasing 25
   5.3   Amplitude Mapping (Time Domain) 25
   5.4   Frequency Mapping (Frequency Domain) . . . 26
   5.5   Map Construction (Spatial Domain) 31
   5.6   Map Features 34
   5.7   Mapping of Evoked Potentials 35
   5.7.1 Latency and Amplitude Determination for EPs
         and ERPs 35

6 Storing of Data 37

7 Statistical Procedures 38

8 Practical Application: Findings in Normal Subjects . . 41
   8.1   Introduction 41
   8.2   EEG Features in the Time Domain 41
   8.2.1 Dipole Estimation 41
   8.3   EEG Features in the Frequency Domain 44
   8.4   EP Features 45
   8.4.1 Mapping of Visual Evoked Potentials 45
   8.4.2 Mapping of Auditory Evoked Potentials . . . 46
   8.4.3 Mapping of Somatosensory Evoked Potentials 49
   8.4.4 Mapping of Contingent Negative Variation
         (CNV) and in Response to Olfactory and
         Chemosensory Stimulation 49

8.5   EEG Mapping After Sensory, Motor,
         and Mental Activation and due to
         Psychotherapeutic Interventions   50
8.6   Sleep Features 50

9 Findings in Diseases   54
9.1   Introduction   54
9.2   Evaluation of EEG and EP Maps   54
9.3   Clinical Examples   56
9.3.1   Introduction   56
9.3.2   EEG Mapping of Local Frequency
         and Amplitude Differences   56
9.3.2.1 States Causing Increased Intracranial Pressure
         (Brain Tumors)   56
9.3.2.2 Cerebrovascular Diseases   61
9.3.3   EEG Mapping of Transients   62
9.4   EEG and EP Mapping During Normal Aging .   62
9.4.1   Changes in EEG Topography   62
9.4.2   Changes in P300 Topography   62
9.5   EEG and P300 Topography in Dementia
         of Alzheimer Type   64
9.5.1   Stage-Dependent Alterations of EEG
         and P300 Mapping in Dementia
         of Alzheimer Type   67
9.5.2   Differential Diagnosis of Dementia   67
9.5.2.1 Luetic Infection (Progressive Paralysis)   69
9.5.2.2 Pick's Disease 69
9.5.2.3 Wilson's Disease 69
9.5.2.4 Parkinson's Disease,
         Parkinson's Disease with Dementia,
         Dementia of Alzheimer Type,
         and Major Depressive Disorder   70
9.5.2.5 Dementia of Alzheimer Type
         and Multi-infarct Dementia   72
9.6   EEG and EP Mapping in Psychoses   72
9.6.1   Case Studies 72
9.6.1.1 Schizoaffective Disorder (DSM-III: 295.7) . . .   72
9.6.1.2 Schizophrenic Disorder, Paranoid Subtype
         (DSM-III: 295.3)   74
9.6.1.3 Major Depressive Disorder (DSM-III: 296.2) .   76
9.6.2   Group Results   78
9.6.2.1 EEG Mapping in Schizophrenia   78
9.6.2.2 P300 Mapping in Schizophrenia   80
9.6.2.3 EEG and EP Mapping in Depression   80
9.7   EEG and EP Mapping in Clinical
         Psychopharmacology   83
9.7.1   EEG Mapping After Application of Drugs . .   83
9.7.2   EP Mapping After Administration of Drugs .   86

10 Advanced Methods   89
    10.1   Dipole Source Estimation   89
    10.2   Neurometrics   90
    10.3   Determining Differences Between Maps   90

References   91

Subject Index   101

1 Introduction

Advances in computer technology and software have made it possible for medical imaging techniques to be developed in the past 10 years that permit the visualization of structures and functional processes of the human brain (Freeman and Maurer 1989 b). The term "neuroimaging" refers to any of a number of procedures for visualizing features of the central nervous system. Imaging procedures that depict structures are computed tomography (CT) and magnetic resonance imaging (MRI), while procedures demonstrating functions include positron emission tomography (PET), cerebral blood flow analysis (CBF, also measured by single photon emission computed tomography, SPECT), magnetoencephalography (MEG), and computerized electroencephalographic topography (CET). The last-named procedure is generally called mapping or brain mapping. It is noninvasive, permitting follow-up examinations to be performed as often as needed, and has extremely short analysis times (in the range of milliseconds). Electroencephalographic (EEG) and evoked potential (EP) mapping do not portray anatomic structures but the constantly varying spatial distribution of the electrical fields generated by the brain.

EEG mapping has become a popular technique since its method and clinical applications were first described (Harner and Ostergren 1978; Duffy et al. 1979). Due to its numerous applications, however, there is a danger of its improper use and evaluation, or even of its misuse (Kahn et al. 1988; Nuwer 1989), and recommendations have been drafted to prevent them (American Electroencephalographic Society 1987; Nuwer 1988 b, c; Duffy and Maurer 1989; Herrmann et al. 1989). The present atlas is intended to be a concise introduction to the recording, storage, automated monitoring, and color display of EEG and EP data, artifact removal, and analysis of EEG and EP data in time, frequency and spatial domains. After that a comparison of EEG and EP data of patients to those of controls and to the values expected for particular disease categories will be done. The views expressed are our own, yet we have incorporated the mapping guidelines mentioned above and suggestions made by many experienced practitioners of mapping (Duffy et al. 1979, 1981; Coppola et al. 1982; Etevenon and Gaches 1984; Walter et al. 1984; Maurer and Dierks 1987 a, b; Borg et al.

1988; Jerrett and Corsak 1988; Nuwer 1988 b, c; Bickford 1989; Desmedt and Chalklin 1989, Fisch et al. 1989; Freeman and Maurer 1989 a, b; Garber et al. 1989; Hughes and Miller 1989; Kohrmann et al. 1989; Lehmann 1971, 1989).

The atlas is not intended to be an exhaustive description but as a reminder of the questions that should be kept in mind as this field expands. We agree wholeheartedly with Nuwer (1988 b, c) that "a thorough familiarity with traditional EEG is a prerequisite to understanding the meaning of the quantitative EEG results and mapping". Maps should only be interpreted in association with the traditional paper tracings of raw EEG data on which quantitative analysis and mapping are performed.

2 History

In 1924 Berger performed the first electroencephalographic recordings on humans (Berger 1929). Initially very few cerebral points served as the source for measurements of electrical activity of the brain. The first step torward mapping was made with simultaneous recording from several points on the skull and Adrian's discovery of the occipitoparietally centered alpha focus and the travelling wave character of the alpha rhythm (Adrian and Yamagiwa 1935).

In 1944 Motokawa succeeded in preparing the first EEG map (Fig. 1). In the 1950s several groups devised approaches for displaying dynamic patterns. These included the betatron of Lilly and the toposcope by Walter and Shipton. Lilly (1950) used a 5×5 array with a specially designed 25-channel amplifier. The

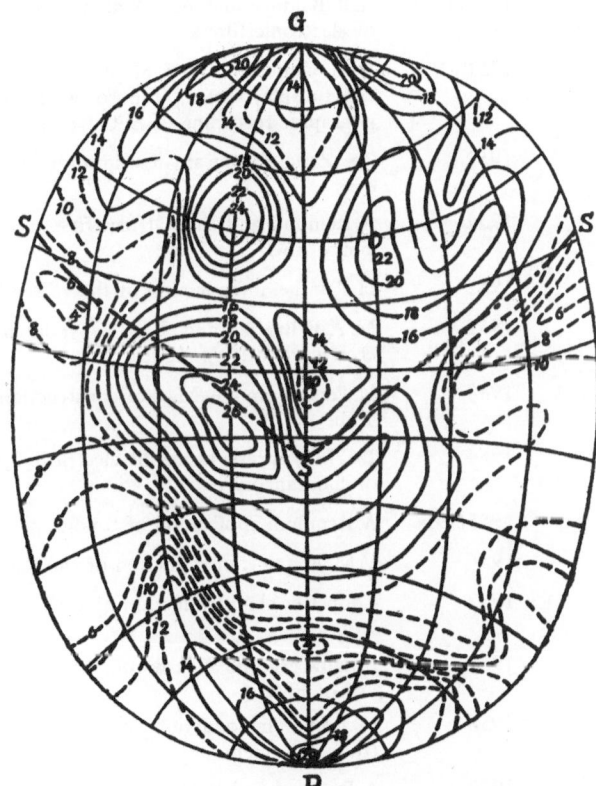

Fig. 1. First EEG map in man showing lines of equal amplitudes in microvolt (µV). G, glabella; S, sulcus centralis; P, protuberantia occipitalis externa. (From Motokawa 1944)

Table 1. History of quantitative EEG and mapping

1875	R. Caton: first tracing of EPs in animal experiments
1924	H. Berger: first EEG recording performed on a human subject
1929	H. Berger: first publication in *Archiv für Psychiatrie und Nervenheilkunde*
1930–1938	H. Berger: 13 further publications about the human EEG
1932	G. Dietsch: first application of the Fourier analysis on the human EEG
1934	E. D. Adrian and A. Matthews: acknowledgement of Berger's discoveries and scientific work
1936	A. Chweitzer et al.: computing instantaneous amplitude histograms during a mescalinic intoxication
1937/1938	Z. Drohocki: introduction of new quantitative EEG analyses and a first frequency analysis named *Electrospectrography*
1937/1938	Grass and Gibbs: application of Fourier transform to the analysis of EEG
1942	K. Motokawa: first bioelectric brain map
1943	I. Bertrand and R. S. Lacape: first book on EEG modelling
1944	K. Motokawa: electrical activity and its relationships to cytoarchitectonics of the cortex
1943/1946	G. R. Baldock and W. G. Walter: frequency analysis by the use of electronic filters
1951	W. G. Walter and H. W. Shipton: development of a new toposcopic display system "toposcope" for optically controlled topographic recording of EEG frequency distribution
1952	Zee Zang Zao et al.: description of the electrical field of the eye
1952	A. Remond and F. A. Offner: first topographic studies of occipital EEG
1952	H. Petsche: introduction of the vector EEG method
1952	M. A. B. Brazier and J. U. Casby: introduction of auto- and cross-correlation function
1954	H. Petsche and A. Marko: construction of a photocell toposcope
1954	J. C. Lilly: presentation of electrical figures during responses of spontaneous activity
1955/1956	A. Remond: application of topographical methods of EEG
1956	V. M. Ananiev: introduction of the electroencephaloscope
1957	Cooper et al.: introduction of a 22 channel helical scan toposcope
1957	B. Saltzberg and A. R. Burch: introduction of the period analysis of EEG
1958	H. Jasper: introduction of the ten-twenty system for electrode placement
1959	A. Remond and R. Delarue: proposal of an EEG system using 58 recording electrodes

1960	W. R. Adey et al.: application of Fourier analysis based on fast Fourier transformation (FFT) and phase analysis to EEG signals – beginning of computerized spectral analysis
1960	H. Petsche and S. L. Stumpf: publication of topographic and toposcopic studies of origin of arousal pattern in the rabbit
1961	T. M. Itil: periodic EEG analysis and use for classification of psychopharmacological agents
1962	H. Petsche: presentation of spike and wave propagation by toposcopic methods
1962	G. D. Goff et al.: representation of scalp topographic studies of EPs
1963	N. P. Bechtereva et al.: localization of focal brain lesions by EEG studies, anticipating clinical EEG cartography
1963	D. O. Walter: further development of EEG spectral analysis
1965	J. W. Cooley and J. W. Tukey: introduction of the fast Fourier algorithm
1966	D. W. De Mott: cortical microtoposcopy
1968	I. Mezan et al.: chronotopographic studies on somatosensory EPs
1970–1973	B. Hjorth: development of new quantitative analytic methods, such as source derivation
1971	J. R. Bourne et al.: topographic characteristics of visual EPs
1971	D. Lehmann: multichannel topography of human alpha EEG fields
1972	R. G. Bickford: compressed spectral array
1972	H. Künkel: simultaneous application of the FFT to several EEG channels (up to 16)
1975	S. Ueno et al.: topographic computer display of abnormal EEG activities
1976	H. Künkel: topographic and psychosomatic aspects of spectral EEG analysis
1976/1978	R. N. Harner and K. A. Ostergren: presentation of computerized topographical EEG data
1978	R. A. Ragot and A. Remond: EEG field mapping
1979	F. H. Duffy et al.: brain electrical activity mapping (BEAM)

first EEG toposcope was developed in 1951 by Walter and Shipton. The major advances in the history of mapping are listed in Table 1.

The recent advances in the field have been the topic of several international conferences, some of which have resulted in informative publications on specific topics, such as *Topographic Mapping of Brain Electrical Activity* (Duffy 1986), *Functional Brain Imaging* (Pfurtscheller and Lopez da Silva 1988), *Topographic Brain Mapping of EEG and Evoked Potentials* (Maurer 1989), *Statistics and Topography in Quantitative EEG* (Samson-Dollfus et al. 1989), and a special issue of *Psychiatry Research* on

Imaging of the Brain in Psychiatry. Moreover, a new journal called *Brain Topography* is devoted to this topic, and a new section on *Neuroimaging* has been added to *Psychiatry Research*.

In addition, two societies have been founded that are devoted to the mapping of EEG and evoked potentials and to MEG: the International Society for Brain Electromagnetic Topography (ISBET) and the International Society for Neuroimaging in Psychiatry (ISNIP).

The rapid development of the field, and specifically of computer technology, in recent decades has made powerful tools widely accessible that make it possible for complicated and tedious calculations to be performed quickly. Yet this is a potential source of problems, as was pointed out by Petsche in 1976:

"Computers, paradoxical as it may sound, frequently decrease the efficiency of work, for nothing is more difficult than to sift the chaff from the wheat considering the huge number of data put out by the computer. The search for significant results has been becoming increasingly difficult. I even dare to claim that, since the computer has come to dominate electroencephalography, it has become much more difficult to distinguish efficient from foolish problems. I see only one way to of escaping this danger: namely to keep in mind the physiological and clinically relevant problems and not to be entangled in problems created by the computer".

One of the goals of this book is to confront such issues and thus to contribute to realizing the potential offered by the use of EEG and EP mapping.

3 Definition and Terminology

Mapping of EEG and EP is a spatially orientated procedure for calculating amplitude and frequency patterns on the basis of measurements from a restricted number of electrodes on the head and by subsequent interpolation methods. Data are generally recorded from 16–64 electrodes on the head, and an interpolation algorithm is used to obtain several thousand points. In contrast to structural imaging methods such as CT, maps of EEG and EP are based on a high proportion of calculated, as opposed to measured, values; EEG and EP mapping may thus be regarded as a "pseudoimaging method." Despite this restriction EEG and EP mapping offer a considerable advantage over conventional multilead derivations.

EEG and EP mapping differs from conventional EEG and EP recordings in a number of ways. The characteristic features of conventional EEG and EP recordings are:

a) Temporally oriented measurements
b) Visual determination of waveform and amplitude maxima and minima of EEG and EP
c) Evaluation of latencies of EP peaks and troughs
d) Advantageous temporal resolution
e) Patterns represented in time (i. e., spikes, EPs)
f) Diagnostic criteria are "graphoelements"
g) Two-dimensional source detection by phase reversal

EEG and EP mapping on the other hand are characterized by:

a) Spatially oriented two-dimensional (maps) or three-dimensional (dipole) measurements
b) Inspection of momentary maps of EEG events (spikes, K complexes, slow activity etc.) and EP maps
c) Evaluation of topographical pattern of peaks and troughs
d) Advantageous spatial resolution
e) Patterns represented in space
f) Diagnostic criteria are topographical maps, or "landscapes"
g) Three-dimensional source detection by dipole estimation

Although EEG and EP mapping is the most common used expression, a number of other terms are also currently in use; these include:

- Computerized electroencephalographic topography (CET)
- Brain mapping
- EEG and EP topography
- EEG and EP mapping
- EEG imaging
- Brain electromagnetic topography (BET)
- Brain electric activity mapping (BEAM®; Nicolet Biomedical Instrument Company)

The term "brain mapping", an abreviation of Brain Electrical Activity Mapping, wrongly suggests a connection with anatomical structures; this should be kept in mind when using it.

4 Methodology

4.1 Introduction

EEG mapping makes use of the same methodological procedures as conventional EEG recording; EPs, however are now recorded under multilead conditions. Different compared to conventional EEG and EPs are (a) the transformation of data from the time and frequency domain into the spatial domain by interpolation techniques, (b) various statistical and other measures to evaluate individual and group abnormalities in various diseases, and (c) multilead recordings of EPs. EEG and EP mapping include a multitude of procedures to measure the spontaneous and the stimulus- and event-related functional electrical state of the brain. The various forms of EEG and EP mapping that are in common use and that are described in this atlas are listed in Table 2 on page 10.

4.2 General Conditions

The room in which measurements are made should be sound-proof and electromagnetically shielded. It is important for the subject to be in a comfortable position to avoid muscle artifacts especially in the shoulders, neck and eyes. During the placement of electrodes patients can be informed about the meaning of the test. The state of alertness should be monitored during the whole session. Special testing, such as that of activated EEG, or task-relevant EP procedures, such as a P300 recording, require such an explanation as well as a test trial to guarantee understanding of the task.

Optimally, patients should be free of medication to prevent drug-dependent alterations in the EEG; if premedication has been administered, exact documentation is necessary, including of dosage. Functional imaging procedures require exact diagnostic classification according to internationally accepted systems (e. g., ICD, RDC, DSM) and a documentation of neurological and psychopathological findings (e. g., Brief Cognitive Rating Scale, Scale for Assessment of Negative Symptoms) and right/left-handedness (Lai 1986).

Table 2. Forms of computerized EEG/EP mapping

EEG mapping in the resting state (awake) with eyes open or closed

EEG amplitude mapping (time domain): presentation of voltage parameters and their topography without FFT
  Amplitudes of EEG waves (delta, theta, alpha, beta)
  Amplitudes of EEG pattern (e. g., spikes/wave complexes, sharp waves, K complexes, sigma spindles)

EEG frequency mapping (frequency domain): presentation of frequency parameters and their topography after FFT
  Spectra of EEG background activity (delta, theta, alpha, beta), presented as
  a) bandpower (μV^2)
  b) square root of power (activity in μV)
  c) relative activity (percentage)
  d) coherence function

EEG mapping in the activated state (dynamic EEG cartography)

Examples in the eyes-open state
  Reading silently
  Reading aloud
  Studying figures, etc.

Examples in the eyes-closed state
  Finger movements (e. g., open and close fist)
  Listening to speech
  Listening to music
  Calculations (e. g., 100 minus 7), etc.

EEG mapping during sleep

EP mapping
EP mapping (exogenous components)
  Mapping of visual EPs
  Mapping of auditory EPs
  Mapping of somatosensory EPs

EP mapping (endogenous components)
  Mapping of auditory, visual, and somatosensory P300
  Mapping of contingent negative variation
  Mapping of slow wave

Mapping of statistical parameters
  Mapping of Z values
  Mapping of t and p values
  Mapping of correlations between parameters
  Neurometrics

According to recommendations of the German EEG Society for the mapping of EEG and EP (Hermann et al. 1989) and to guidelines proposed by Duffy and Maurer (1989), standardized and comparable recording and assessment conditions are required. In the course of a routine EEG maps of the following standard conditions should be obtained:

20 s in the eyes closed state
20 s in the eyes open state
20 s during the last 20 s of hyperventilation (between 2 min 40 s and 3 min).

In addition, the clinician or researcher can record data and maps of EEG and EP of his own choice.

4.3 Calibration

A complete calibration of the system must be performed at the start of each recording. A minimal requirement is the ability to apply a known single-frequency sine wave of calibrated amplitude (e. g., 10 Hz and 100 µV peak to peak) at the front of each amplifier. Additionally, measurements of distortion, phase differential, cross-talk, and bandpass limits are desirable. Bandpass 3 dB points, roll-off, noise figure, and filter type should be specified for all amplifiers.

4.4 Electrodes

Electrodes are placed on the head according to the 10–20 system (Jasper 1958). The 10–20 array is the basis for EEG and EP mapping as this guarantees internationally comparable results. In the classical 10–20 system, 19 to 20 electrodes can be placed (Fig. 2), with the distance between electrodes some 5–6 cm depending on the size of the head (Eppstein and Brickley 1985). Various procedures have recently been proposed to extent the

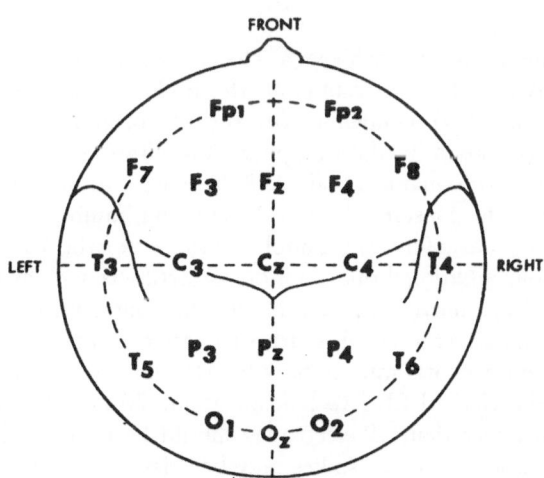

Fig. 2. International 10–20 system of electrode placement, with the alphanumeric designations of electrode placements on the scalp for EEG recordings as used in this atlas

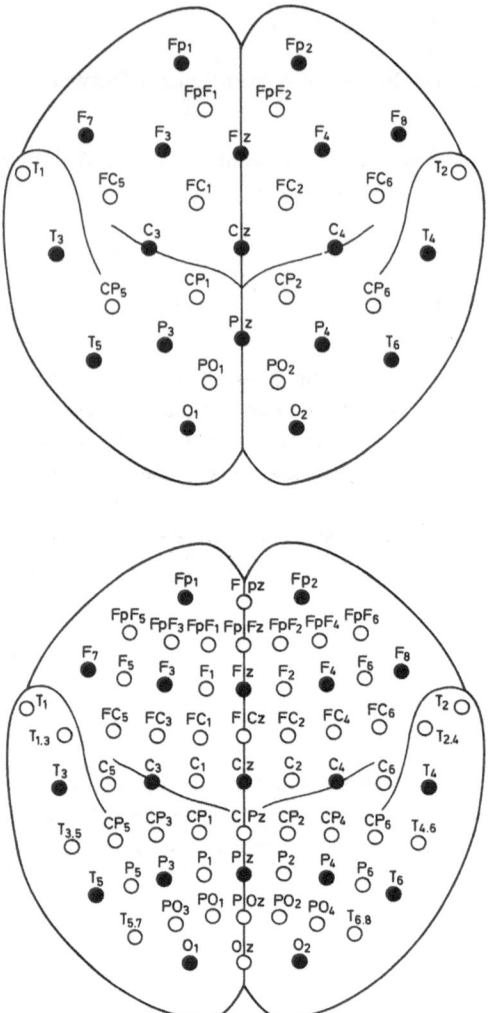

Fig. 3. Recommended electrode placement with a multichannel mapping unit. Electrodes according to the 10–20 system *(filled circles)* supplemented by positioning the 10% system *(empty circles)*. (From Herrmann et al. 1989)

Fig. 4. The 10% system for electrode placement with designations recommended by the German EEG Society. (From Herrmann et al. 1989)

10–20 system and to achieve a closer spacing of electrodes (Fig. 3; Nuwer 1987; Chatrian et al. 1988; Herrmann et al. 1989; Spitzer et al. 1989). Guidelines for the use of more electrodes have been proposed by the Mapping Committee of the German EEG Society (Hermann et al. 1989), where a maximum of 67 electrodes is considered (Fig. 4). The minimum number of electrodes for basic screening should be the 19 of the 10–20 system plus four additional ones to monitor artifacts such as muscle activity and eye movements. Optimally the four additional electrodes should cover vertical and horizontal eye movements and temporal and occipital muscle activity. Affixing the 23 electrodes takes an experienced EEG technician about 20 min. Recordings made with fewer than 19 electrodes should be rejected, for important transients such as spikes may be missed (Fig. 5).

Fig. 5 a–c. Spike activity recorded with 4 (a), 10 (b) and
19 electrodes (c). Only the
19-electrode setting detected
the negative spike at C4

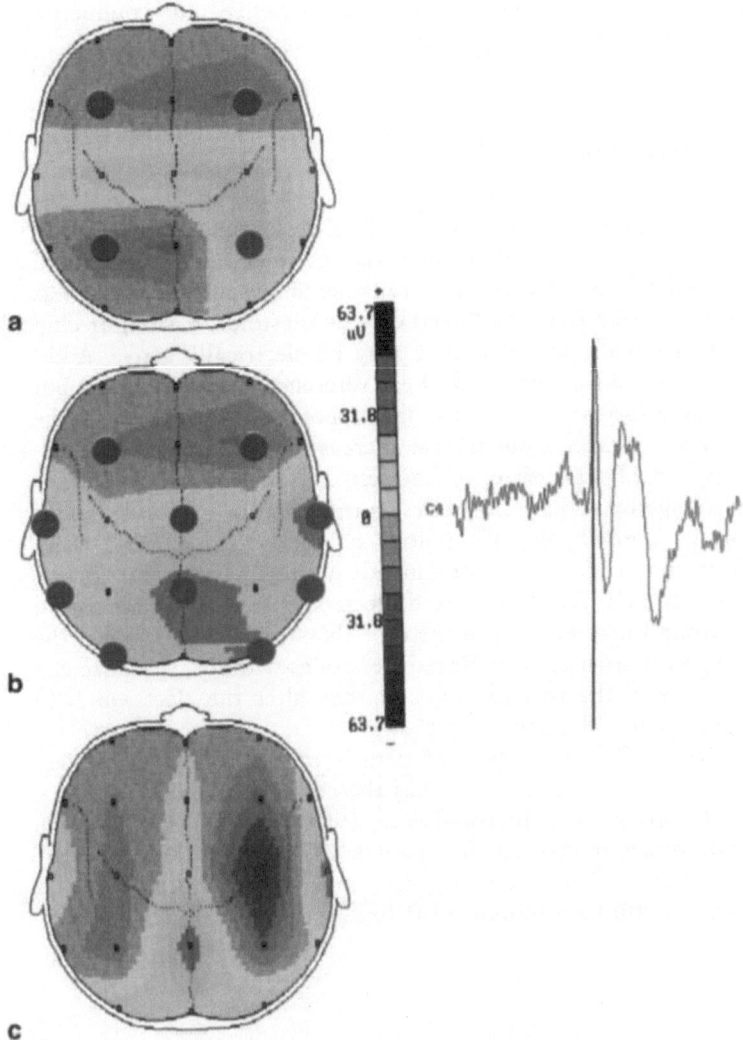

There is a trend toward multilead recordings with 32 electrodes. In this case the electrode array shown in Fig. 3 is advisable. The number of electrodes is limited in the clinical setting by
the need to complete electrode placement in 20–30 min. The
maximum number of electrodes used at present is 128 (Gevins
1989). This high number of electrodes, allows spacing with
2–2.5 cm between electrodes compared to 5–6 cm in the
10–20 system using 19 electrodes. Gevins (1989) reported that
four technicians require approximately 45 min to fix 128 electrodes to the skull, which makes it almost impossible to use this
procedure in a clinical setting.

The same sort of electrodes can be used for EEG and EP
mapping as for a conventional EEG recording. However disk
electrodes are advisable that have a diameter of some 3–5 mm,

and that can be mounted using self-adhesive and conductive paste. These provide long-term stability and low resistance.

4.5 References

With mapping the problem of reference becomes more salient. Controversy still surrounds the choice of reference locations, and no ideal solution is currently available. Single references such as ear or mastoid (A1/A2), linked ears or mastoids, nasion, or chin suffer from the fact that they may be electrically active. Additionally for an ear or a linked ear reference, there is a small but noticable increase in activity and a potential difference as the electrode distance from the ear increases. In measuring EEG activity and EPs this effect is manifested as an increase in activity or voltage of midline electrodes as opposed to temporal regions (Fig. 6). The Mapping Committee of the German EEG Society (Hermann et al. 1989) recommends a single electrode as reference (e. g., Cz, A1, A2). Linked electrodes (e. g., ear or mastoid electrodes) may be used in the case that resistances between the two points are used. Two linked electrodes without resistance can short-circuit the two regions and thus alter the electrical field (Katznelson 1981; Lehmann 1987).

Before further analysis of data is carried out, more or less reference-free techniques such as the common average reference (CAR; Offner 1959; Bertrand et al. 1985; Lehmann 1987) or the local average reference (LAR; Hjorth 1975, 1980) should be calculated for spectrum and voltage computation. Both CAR and LAR can also be determined off line.

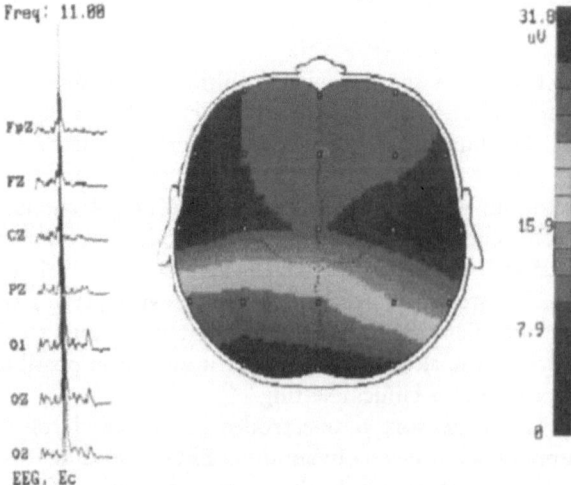

Fig. 6. Map of alpha activity at a frequency of 11.0 Hz, reference link mastrids; noticeable is an increase in activity with a midline maximum along transversal electrodes

Fig. 7. Schematic drawing showing an EP recording obtained at electrode sites Oz (30 μV), Pz (40 μV), Cz (20 μV) and Fz (10 μV). Values for the common average reference are calculated by subtracting the mean of all amplitude values (25 μV) from originally measured amplitude data

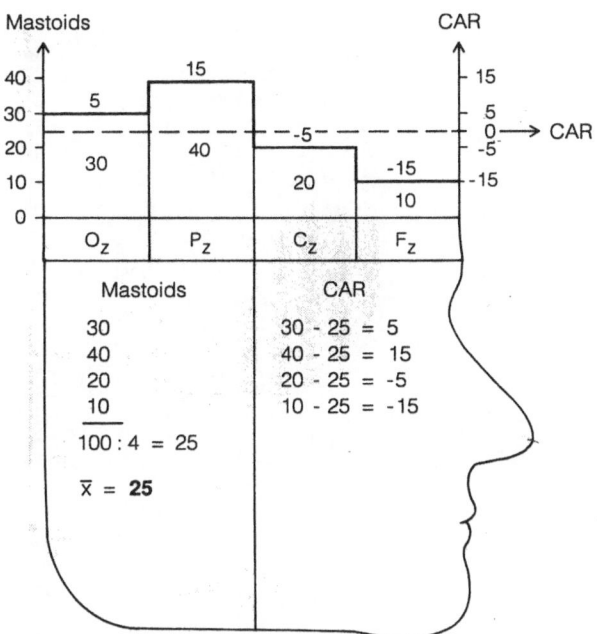

Fig. 8. One of the procedures for computing EEG parameters with the Laplacian transformation. The average potential between C_z and the nearest four electrodes (3.75) was calculated in this example by subtracting surrounding potential values from the C_z-value followed by a division through 4.

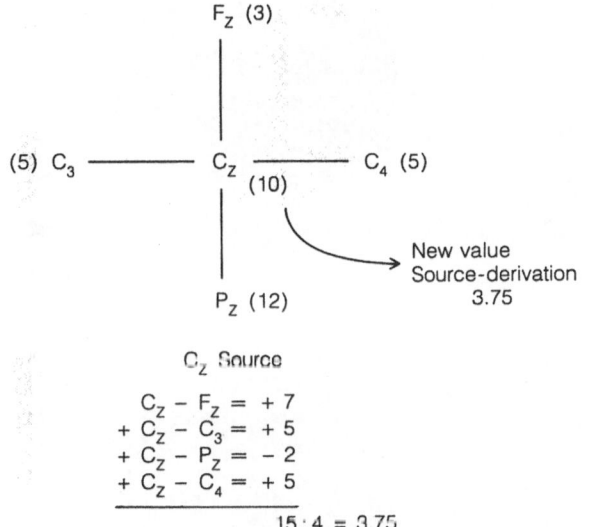

CAR is the difference at a defined point in time between the activity measured at one electrode and the average of all electrodes. Figure 7 presents an example of this during a P300 recording using, for didactical purposes, only four channels. The mean value of the four channels at 330 ms was 25 μV. Thereafter the differences between original data and the average voltage were calculated and regarded as new values for these four channels.

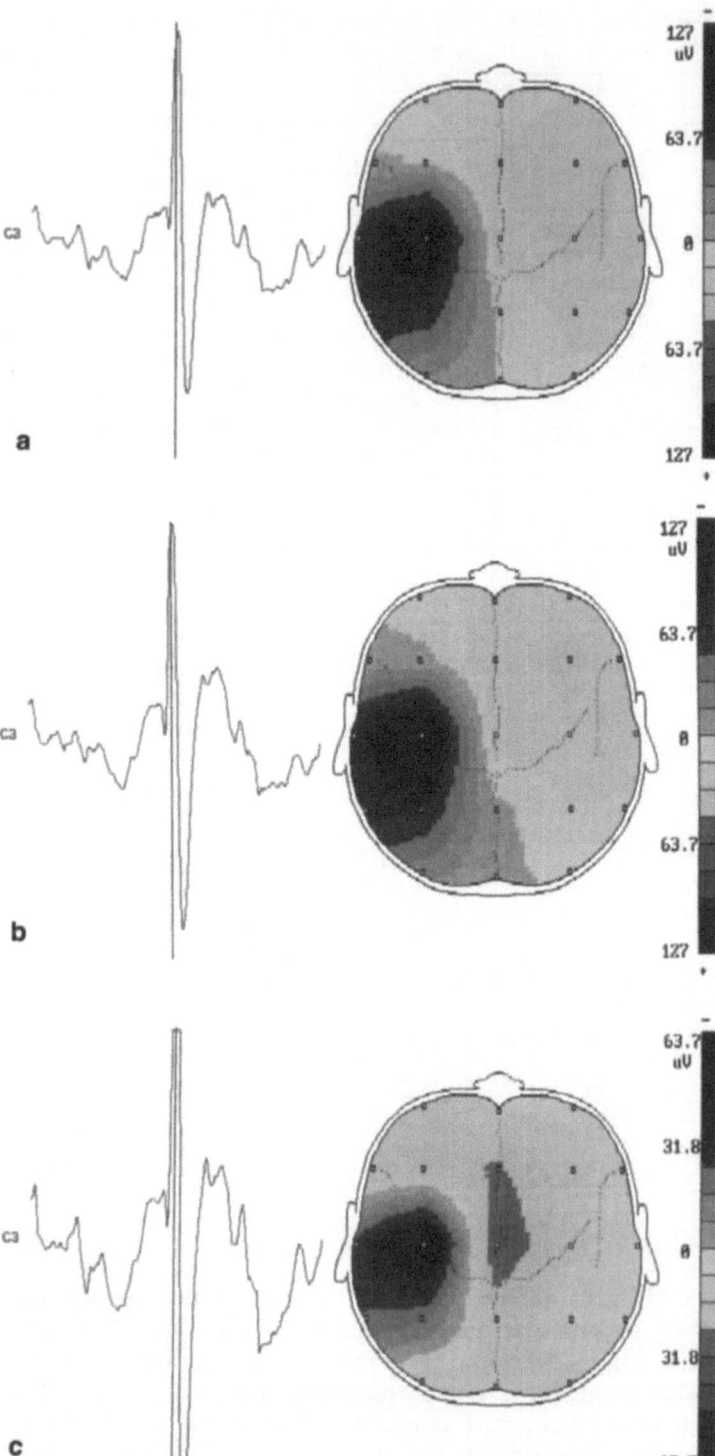

Fig. 9 a–c. Topography of a negative spike recorded against linked mastoids (a), common average reference (b), and current source estimation (c)

Fig. 10. Schematic drawing showing an EEG recorded at midline electrodes Oz, Pz, Cz, Fz, and Fpz in the time domain and transformed to the frequency domain. After a change in reference into a CAR alpha activity, which was preserved at occipital sites, occurred at Fz and Fpz

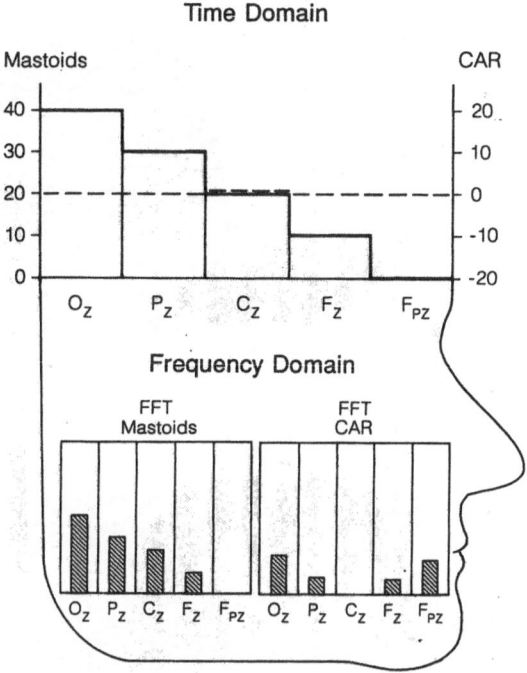

LAR, also known as source derivation (Hjorth 1975) or Laplacian transformation, is another reference-free calculation method that uses the Laplacian law and computes the average potential difference between the nearest four electrodes for each channel (Fig. 8). This method estimates the density of local radial currents and can be interpreted as a spatial high pass filter which emphasizes shallow cortical generators. There are numerous versions of the surface Laplacian approach (Nunez 1988). A disadvantage of LAR is the requirement of equal distances between the electrodes; also, with marginal electrodes only a 3-point operator can be used. LAR in contrast to CAR changes original recorded landscapes. Figure 9 compares (a) an EEG-transient (spike) recorded with linked ears as reference, (b) further data analysis with the calculation of a CAR, and (c) a current source estimation (LAR).

When references are changed, there is no alteration in topographical shape in the time domain and the landscape is maintained for EPs, spikes, K complexes, and other factors; there is a change only in the zero-labeling equipotential line (reference altitude; Fig. 10). In the frequency domain, however, the choice of reference does influence maps. Depending upon the reference used, activity seems to occur at different locations.

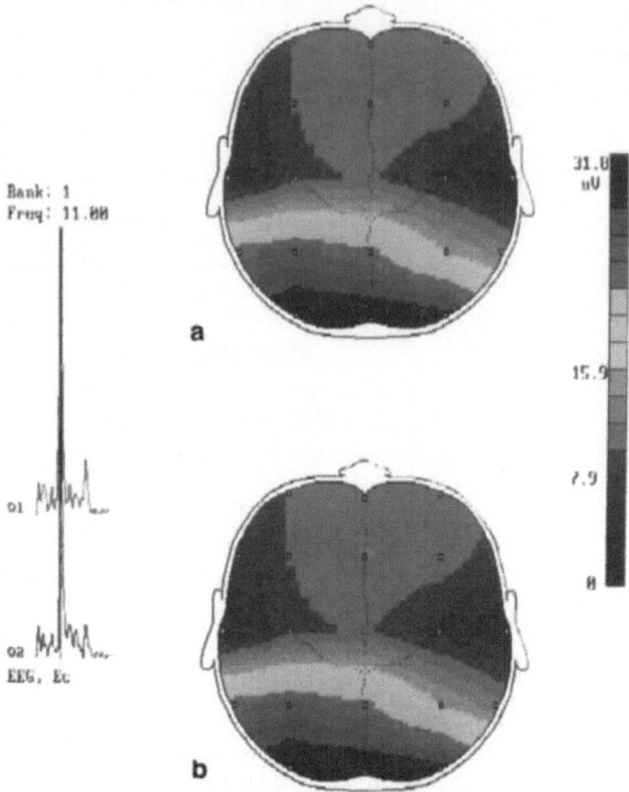

4.6 Baseline

The baseline for the EEG and EPs is the technical zero line (i. e.,
zero voltage difference between inputs of amplifiers). Whereas
the pattern (landscape) of an EEG in the frequency domain after
fast Fourier transformation depends upon the choice of reference electrode and not upon the baseline, the pattern of EP maps
or spikes (time domain) depends on the baseline (Fig. 11 a–d).

4.7 Artifacts

Artifacts in EEG and EP recordings occur frequently and their
removal is the most important task of technicans (Lee and
Buchsbaum 1987; Nuwer and Jordan 1987; Coburn and Moreno 1988). Table 3 presents common biological and technical artifacts. Prior to submission for FFT and EP analysis and mapping,
the incoming EEG should be screened for artifacts and affected
segments removed. The removal is at present best done by the

Fig. 11 a–d. Maps under different baseline conditions. After changement of baseline in one channel no change in pattern could be observed for an alpha FFT (**a** and **b**; frequency domain); evident change in pattern, however, occurred for an alpha-amplitude map (**c** and **d**; time domain)

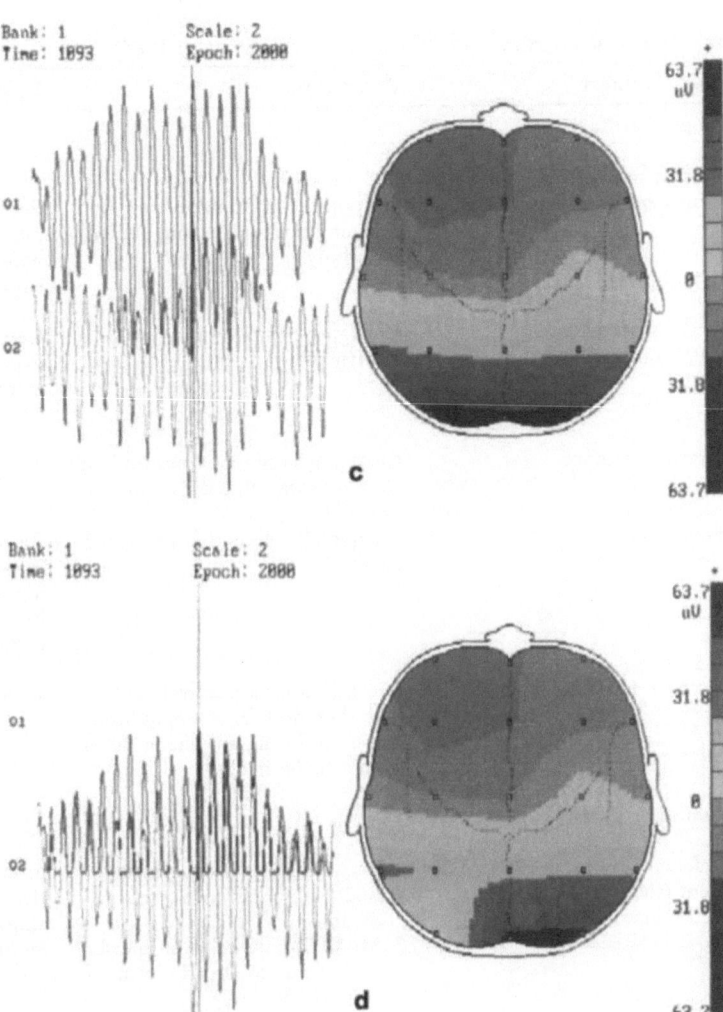

human eye in the case of EEG. Since EP data are more difficult to screen for artifacts due to the averaging process, a system should be provided which allows a single trial elimination on the basis of overvoltage criteria. For proper artifact recognition additional channels to detect eye blinks, horizontal and vertical eye movements, and muscle and movement artifacts should be available (Fig. 12; Irrgang and Höller 1981). Common eye and muscle artifacts and their topographical patterns are shown in Fig. 13 a, b. Some computer software systems are sufficiently sophisticated to detect eye movements. Delta activity sharply localized to the frontopolar region on maps should be presumed to be of ocular origin unless proven otherwise.

Table 3. Common artifacts in EEG (from Pichlmayr et al. 1984)

Cause of artifact	Avoided by	Remarks
Biological artifacts		
Muscular movements (blinking, eye movements, muscular tremor, etc.)	Talking with the patient; patient adopting comfortable sitting or reclining position; asking patient to lie calmly	Artifacts due to movements usually disappear after induction of anesthesia and muscle relaxation. More pronounced movements may simulate activities in the delta band
Cardiac artifacts (R peak in EEG)	Changing position of reference electrode	Cardiac artifacts are of little relevance if recognized early. In spectral analysis a peak of 1–2 Hz may be seen
Pulse-wave artifacts	Changing position of electrode if electrode was placed over artery	
Perspiration	Adequately ventilating recording room.	Respiratory artifacts are very slow potentials, suppressible by changing the time constant, leading to a loss of information in the delta band
Electrodes and cables	Decreasing electrode resistance (less than 50 kΩ). Keeping head still even during anesthesia (head should be limited)	Transition of biological to technical artifacts
Technical artifacts		
Interference from 50/60-Hz AC current (electric mains): "electric buzzing"	1. Excluding faulty equipment 2. Reducing impedance of electrodes 3. Modifying the room if interference is due to electric or magnetic fields 4. Separating cables for recording and power supply 5. Compensating different potentials, when several electric sets are used on one patient	In modern equipment with high electrode impedance, buzzing can almost always be controlled by a 50-Hz filter
High-frequency artifacts, e. g., due to cautery, neon lamps	Removing the cauterant, usually impractical. Replacing tube or transformator. Screening all electric lights	Caution: during cauterization, EEG recorder "bloc" position, otherwise recording pens may get damaged
Cable artifacts through general disturbance in the recording room	Reducing number of people in recording area; relaying cables in protected areas	Reducing the number of people is not always possible

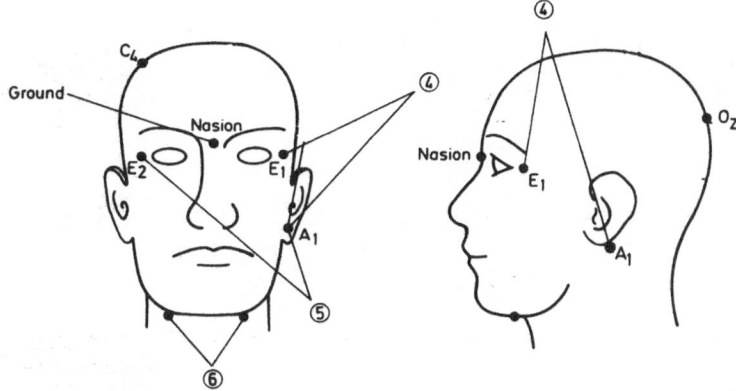

Fig. 12. Electrode placement of E1, E2 and A1 for detection of eye blinks and muscle activity. (From Nicholson and Martin 1983)

A physiological phenomenon and not an artifact, but crucial in a similar way, is drowsiness. Symmetrical differences in slow activity occurring suddenly out of the background or faster activity must be presumed to be related to drowsiness.

It is also the technician's responsibility to reduce the occurrence of artifacts (Rodin 1988). This can be achieved most effectively by placing the patient in a comfortable position, with arm and leg muscles relaxed. In the eyes-closed state, eye blinks can be reduced by covering the eyes with wet padding; in the eyes open-state, providing a point upon which vision is to be fixed may help. The mouth should be kept slightly open to avoid grinding of teeth. Duffy (1981) proposed so-called "blink holidays." Finally, it may be helpful to interrupt the recording and to converse with the patient. In cases in which the cooperation of the patient cannot be obtained (e. g., with severely demented patients) a clinically useful recording is not possible.

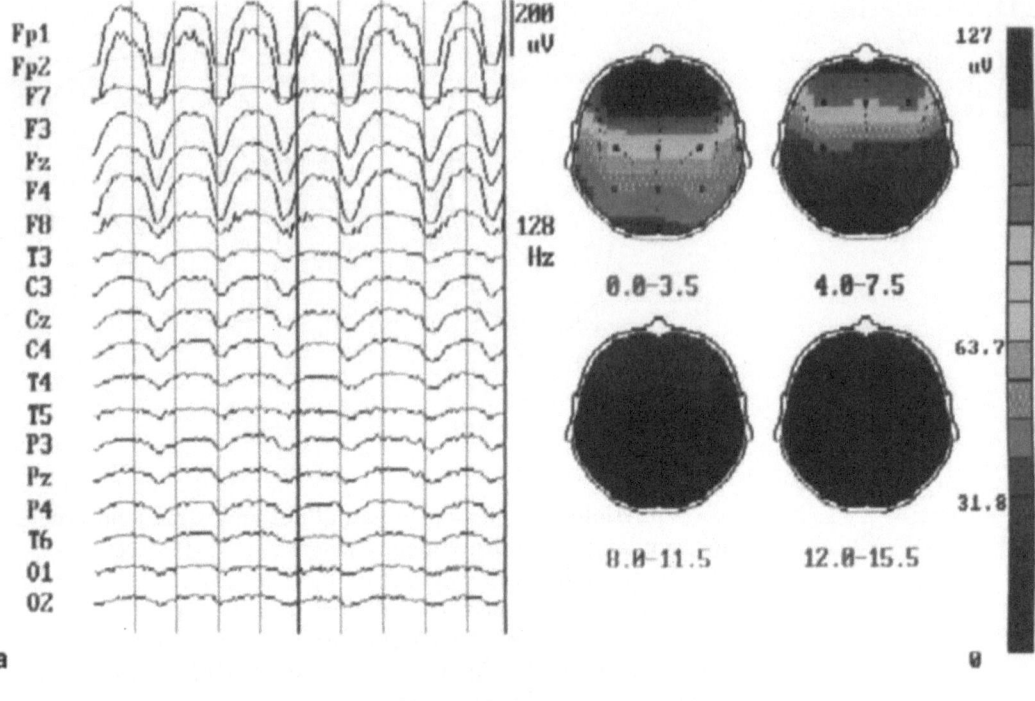

a

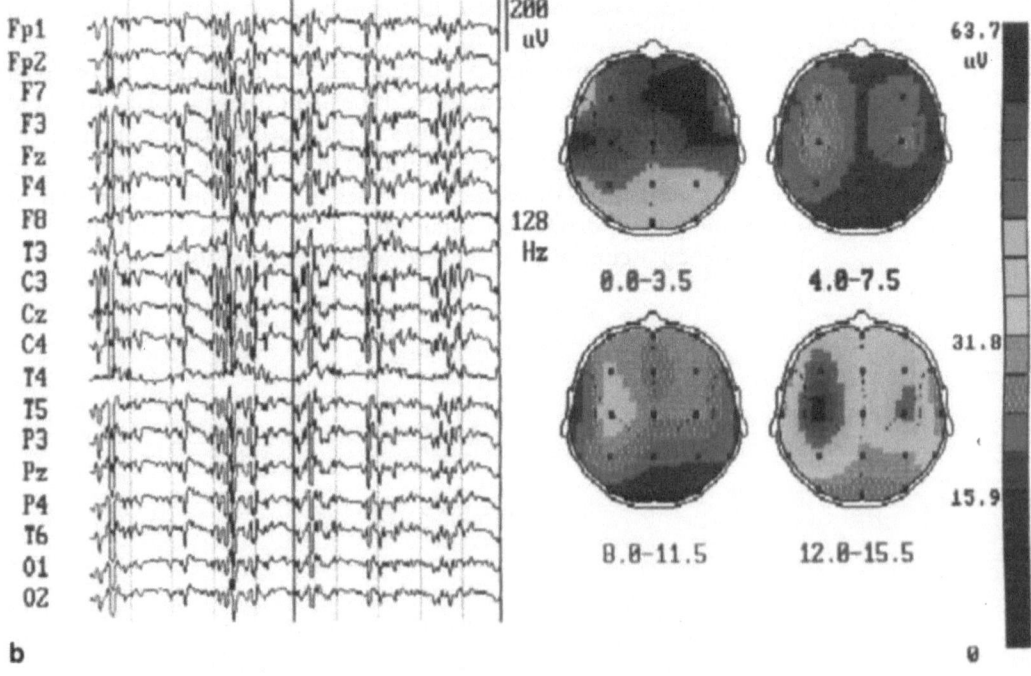

b

5 Data Acquisition and Signal Analysis

Figure 14 demonstrates the path from data acquisition at the scalp electrodes to final map construction on the computer screen.

5.1 Analog to Digital Conversion

EEG and EP signals present continuous variations of potentials as a function of time, however for processing by computers signals must be digitized. The analog to digital conversion (ADC) is the process by which an external analog EEG or EP signal is transformed into the digital equivalent, thus allowing calculations in computers (Fig. 15). The quality of ADC depends upon such characteristics as sampling rate and number of bits of the analog-to-digital converter. In ADC the current signal amplitude of the continuous analog input is measured at regular intervals. The frequency at which the measurements are made is the sampling rate.

The choice of an appropriate sampling rate depends upon the frequency content of the incoming signal. Slow EEG waves require a low sampling rate with relatively long intervals between the measurements of the analog signal, while events with high frequency content (e. g., spikes) require a high sampling rate with short intervals between the measurements. A sampling rate of 60 Hz per channel is the absolute minimum to preserve the EEG information in the range of 1–30 Hz.

The dwell time is defined as the time between two successive digitalized values:

$$\text{Dwell time} = \frac{1}{\text{ADC rate}}$$

The accuracy of ADC depends upon the number of bits with which the analog-to-digital converter resolves the signal. The number of bits determines the dynamic range of the signal (range between the lowest measurable value and the highest, without clipping the signal). Analog-to-digital converters installed in mapping systems generally have a resolution of 8–16 bits.

Fig. 13 a, b. Topographical features of artifacts. **a** Eye movements with high activity (approximately 120 μV) in the delta and theta ranges at electrode sites Fp1, Fp2, F3, Fz, and F4. **b** Movement and muscle activity with an accentuation of delta activity at Fz and F4 and fast activity at C3 and C4

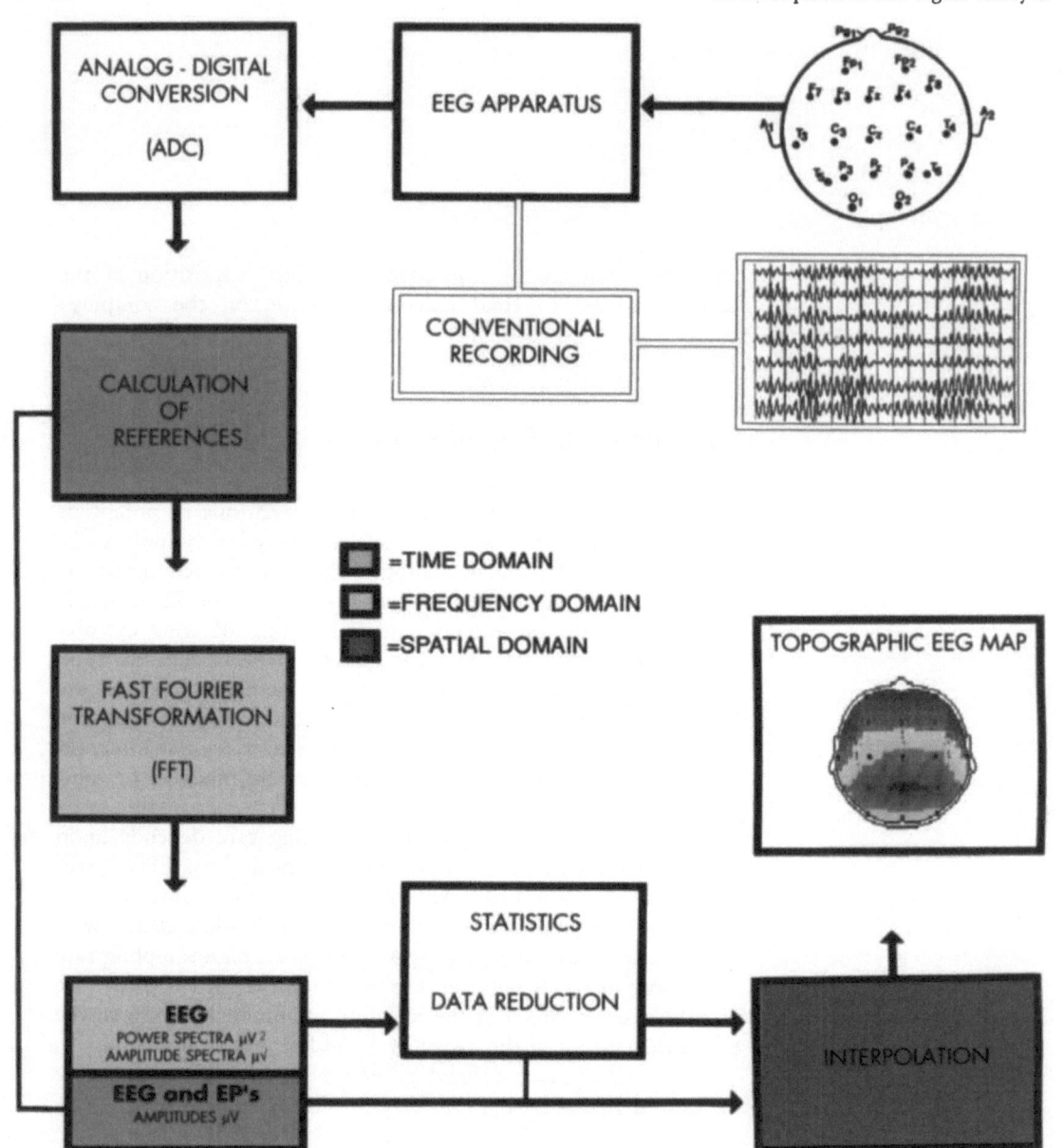

Fig. 14. Schematic diagram
illustrating the data flow from
an EEG recording to frequency or amplitude maps of EEG
and EPs. *Green,* time domain;
yellow, frequency domain;
white, line format display;
red, spatial domain

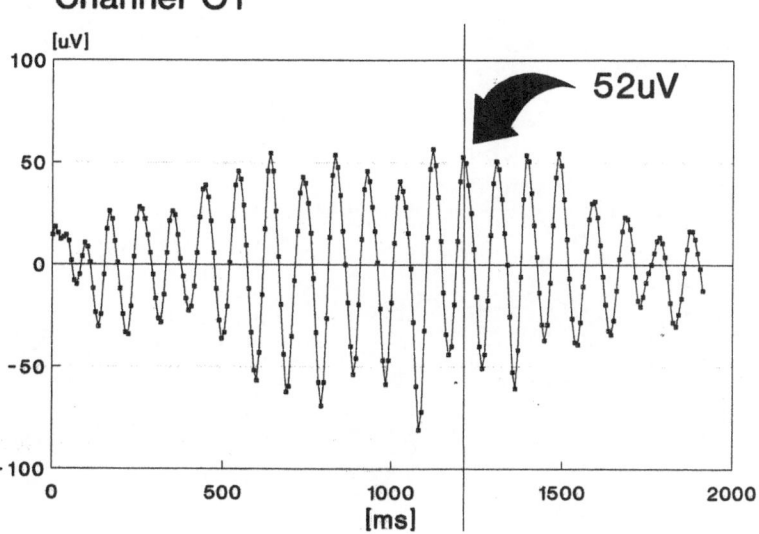

Fig. 14. Schematic diagram
illustrating the data flow from
an EEG recording to frequency or amplitude maps of EEG
and EPs. *Green,* time domain;
yellow, frequency domain;
white, line format display;
red, spatial domain

Fig. 15. Process of analog to
digital conversion. An occipital EEG signal (alpha at O1)
has been digitized with an
sampling rate of 128 Hz

5.2 Aliasing

Aliasing (falsification of the signal) is closely associated with the
sampling rate of the signal. Aliasing occurs when a signal is sam-
pled at a rate that is too low. The Nyquist frequency is the theoretically maximum allowed frequency content of the input signal
and is twice the ADC rate. If the minimum ADC rate is not
achieved, i. e., if the incoming signal contains frequency components greater than the Nyquist frequency (twice the ADC rate)
aliasing occurs, with unpredictable errors in the digital waveform
compared to the analog signal. This may result in altered waveform morphology in the case of EEGs and subsequent inaccurate
peaks in the frequency spectra. Aliasing is an irreparable error
and cannot be corrected once ADC has been completed. It can be
prevented by appropriate filtering of data before ADC.

5.3 Amplitude Mapping (Time Domain)

The EEG can be represented as amplitude versus time, graphical-
ly with time on the x-axis and amplitude on the y-axis. This is
the case in conventional paper EEG records. In the time domain
no FFT takes place. The amplitude values of an EEG at a certain
point in time are mapped. The principal use of amplitude values
is in the cartography of epileptic features, focal disturbances and
display of special EEG patterns such as K complexes, spindle activity, and similar phenomena.

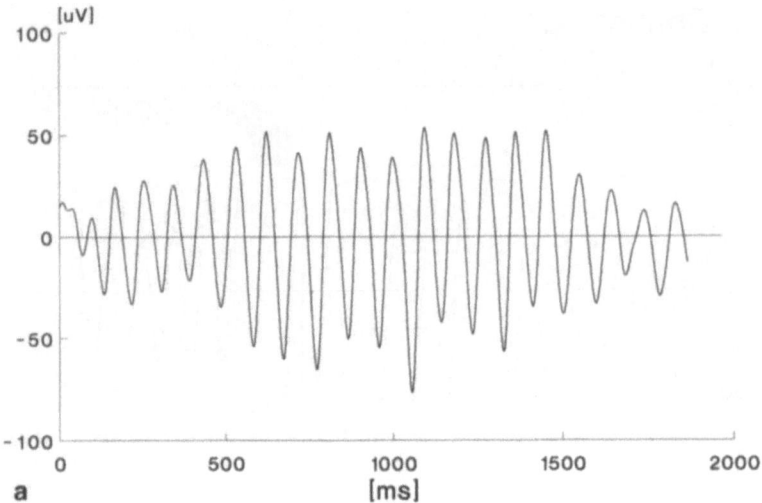

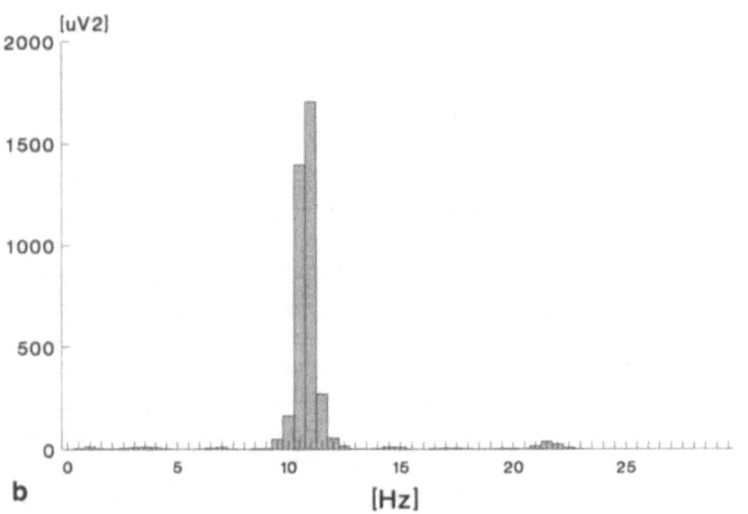

Fig. 16 a–d. Transformation of a 2-s alpha EEG segment from the time domain to the frequency domain. a Original 2-s EEG segment. b Power spectral analysis (μV^2) in discrete frequency bands. c Amplitude spectral analysis (μV) in discrete frequency bands. d Amplitude spectral analysis (μV) in a continuous frequency domain

EPs are represented chiefly in the time domain. EP maps show the amplitude distribution on the scalp at defined latencies or time intervals. The advantage of measuring data in the time domain is the incorporation of information concerning wave shape and latency between events.

5.4 Frequency Mapping (Frequency Domain)

Transformation of data from the time domain into the frequency domain was time consuming and tedious before it was possible to perform mathematical procedures in computers. Cooley and

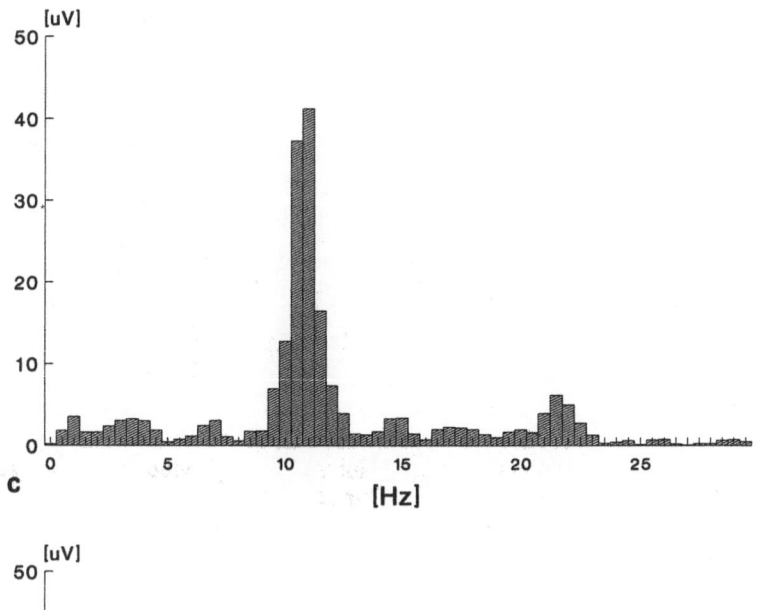

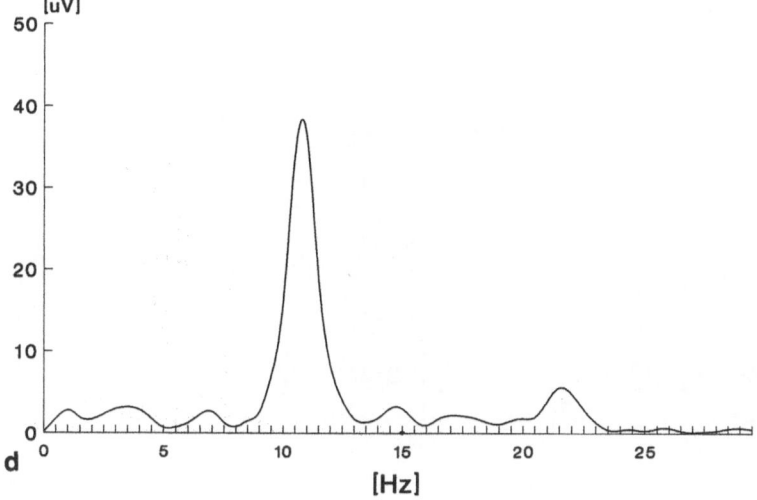

Tukey (1965) developed a special algorithm for Fourier transformation which allowed a faster calculation of frequency spectra (hence, fast Fourier transformation, FFT). In computers the FFT is calculated with special programs or with special microchips that allow it to be calculated even on-line during EEG recording without data loss.

An example of a transformation from the time domain to the frequency domain is shown in Figs. 16 and 17 for a 2-s alpha and a 2-s beta EEG segment. Amplitudes in the frequency domain are represented on the y-axis and frequencies on the x-axis. It is important to know that once data have been converted to power or frequency spectra one can no longer reconstruct the original waveform in the time domain due to a loss of phase in-

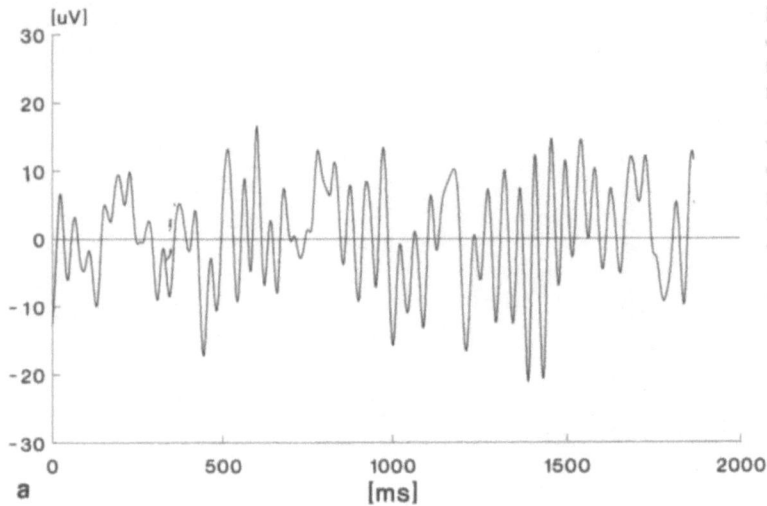

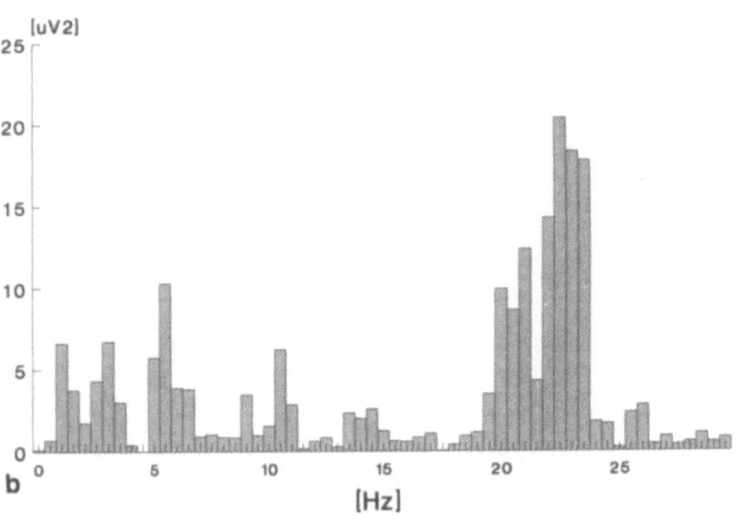

Fig. 17 a–d. Transformation of a 2-s beta EEG segment from the time domain to the frequency domain. **a** Original 2-s EEG segment. **b** Power spectral analysis (μV^2) in discrete frequency bands. **c** Amplitude spectral analysis (μV) in discrete frequency bands. **d** Amplitude spectral analysis (μV) in a continuous frequency domain

formation. EEGs are often dominated by one frequency band, as can be seen in Fig. 16 for alpha and in Fig. 17 for beta. FFT allows a well defined separation of these components. The EEG traces in Figs. 16 and 17 are converted primarily by "power-spectral analysis" into power spectra. Data in the power spectral analysis are measured in terms of square microvolts (μV^2) within a frequency band such as alpha or in terms of square microvolts/ Hertz (μV^2/Hz) when measured along a continuous frequency axis. The disadvantages of power spectra (μV^2) are related to the fact that dominant frequencies are overemphasized due to the mathematical squaring of these values. In this atlas therefore merely maps after amplitude spectral analysis will be shown (square root of power in μV).

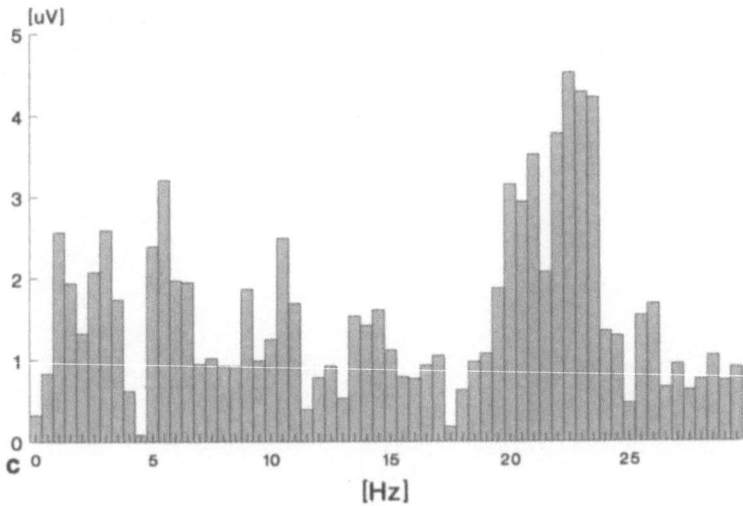

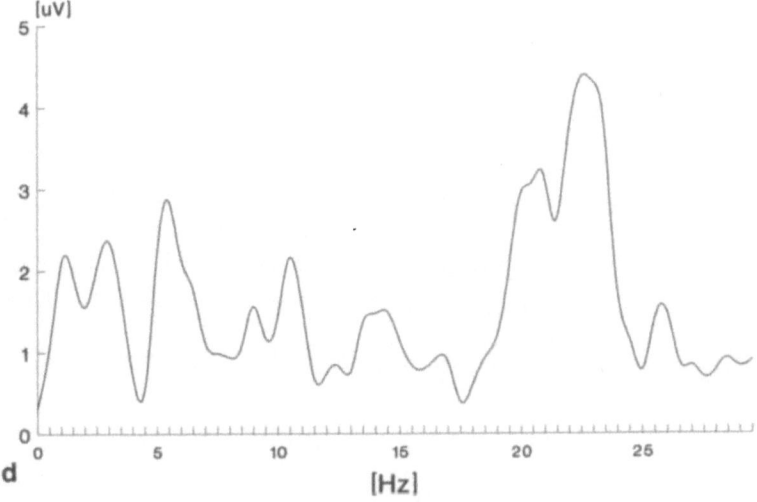

For normalization of data relative amplitude spectra are usually desired, i. e., the percentage of activity for each frequency point of the total activity within the spectrum. The advantage of this method is that in cases in which a person has, for example, a high alpha activity and therefore a high total amplitude and another person low values the relative alpha activity may be similar in the two cases (Fig. 18).

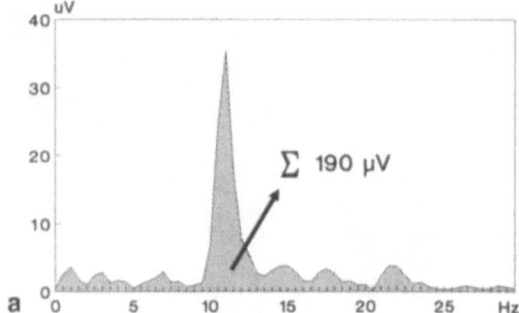

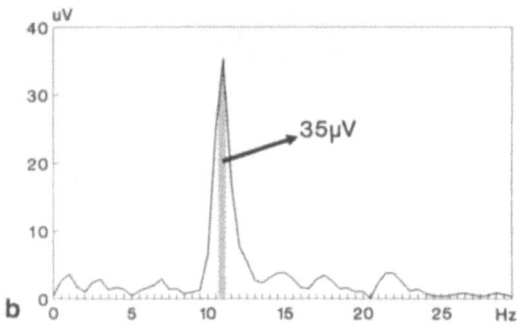

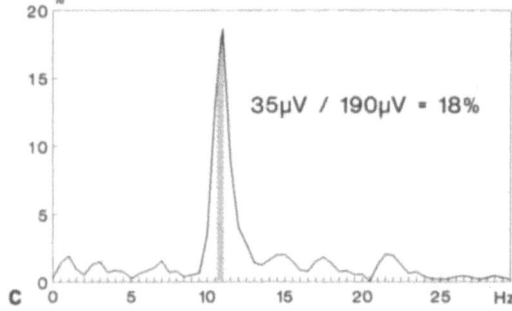

Fig. 18 a–c. Example of a relative amplitude spectrum with a total activity of 190 µV in a frequency range of 0–31 Hz; relative activity for 11 Hz was calculated by dividing the activity at 35 µV by the total activity of 190 µV; the result was a relative activity of 18%

5.5 Map Construction (Spatial Domain)

For map generation an image must be constructed from data measured at only relatively few points on the scalp. Starting with a limited number of actual data points an image consisting of thousands of data points (pixels) is generated. Figure 19 (row 3 and 4) shows that a display of the few actually measured values represents little in the way of a map even after transformation of the values using a color scale. Interpolation is used to fill in the gaps between the 19 values. Linear interpolation uses the three or four nearest electrodes, i. e., a pixel value is treated as the mathematical average of the four nearest electrodes, inversely proportional to the distance from each. Linear interpolation has the advantage of rapid calculation by the computer. A disadvantage is that maxima and minima of activity are always be located at electrode sites. Other interpolation methods, such as surface spline interpolation (Ashida et al. 1984; Perrin et al. 1987), exhibit maxima or minima between electrodes and produce smoother maps, however, the computing time required is considerably longer. The example in Fig. 19 shows an interpolation of values between the electrode sites using the four nearest electrodes and calculating about 4000 numbers or data points (pixels). Confronting such a huge number of values is rather confusing, and only a translation into the spatial domain with gray or color maps can make the information understandable. Figure 19 illustrates how an easily understandable and interpretable map is computed from 19 measurement points on the scalp.

Other methods for presenting the information in a simple and absorbable manner include contour maps (isopotential lines) and three-dimensional head models (Fig. 20). The advantage of three-dimensional head models is their reference independence in the time domain (Dierks et al. 1989 a). Nuwer (1988) introduced bar-format displays (Fig. 21). Compared to line formats these have the advantage of showing the whole time or frequency axis continuously, in contrast to a single map with its representation of data at only one point or band along the time or frequency axis.

The transformation from a restricted number of measured values into a perceivable map by means of interpolation is the main difference to conventional EEG and EP recordings and especially to quantitative EEG.

Maps can now be generated from (a) the ongoing EEG as amplitude presentation (i. e., spike evaluation, K complexes), (b) the spectral content of the EEG, and (c) amplitudes of EPs. In addition, the results of statistical procedures using, for example, Student's t test and z statistics can be visualized as maps. Figures in this atlas include amplitude maps, spectral amplitude maps (square roof of power in μV), maps of EP amplitudes, and statistical maps comparing groups.

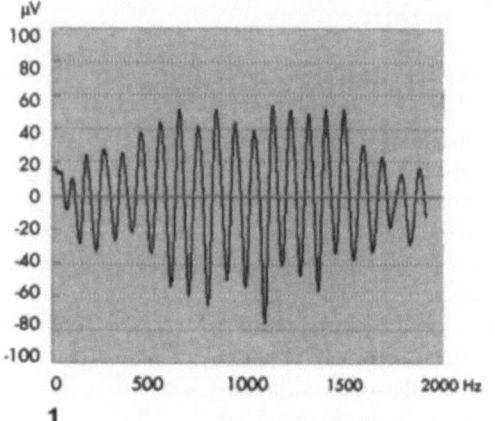

1

Fig. 19. Conversion of EEG data from the time domain to the frequency and spatial domains. After FFT this and the following examples give the square root of power (μV) instead of power μV². *Row 1 and 2:* The maximal amplitudes in the bar diagrams (in the example 41 μV at 11.5 Hz at point O1) correspond approximately to the average amplitude of the same frequency in the original EEG *(row 1). Row 3:* Generation of a topographic EEG map with original EEG *(left),* and 19 frequency spectra *(middle). Row 4:* Map generation using coloration of actually measured values *(left)* and by means of linear interpolation *(middle and right),* where the activity was actually measured at points Fz, F4, Cz, and C4. Calculated by interpolation were 21 intermediate values. *Row 5:* EEG topogram for the alpha frequency (11.5 Hz). For didactical reasons a focus of high alpha activity has been constructed at point F4.

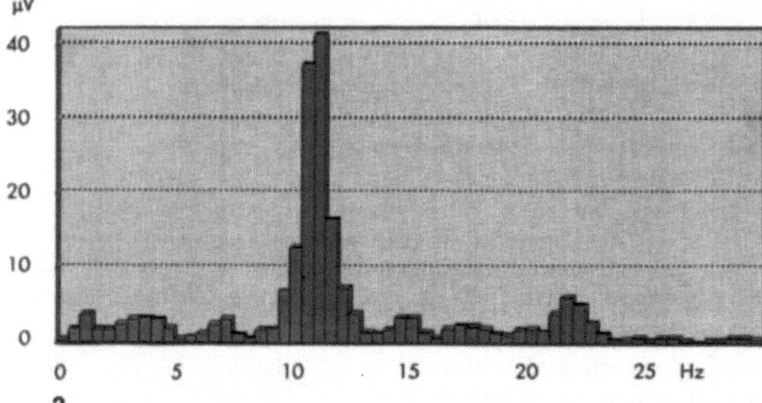

2

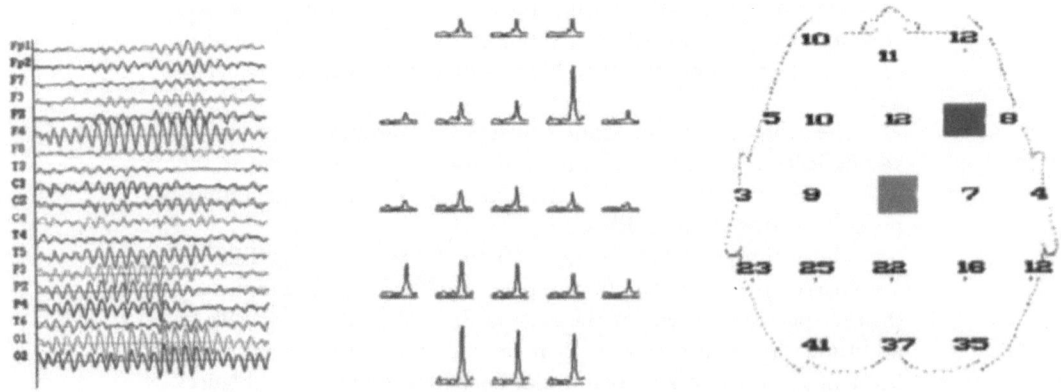

3

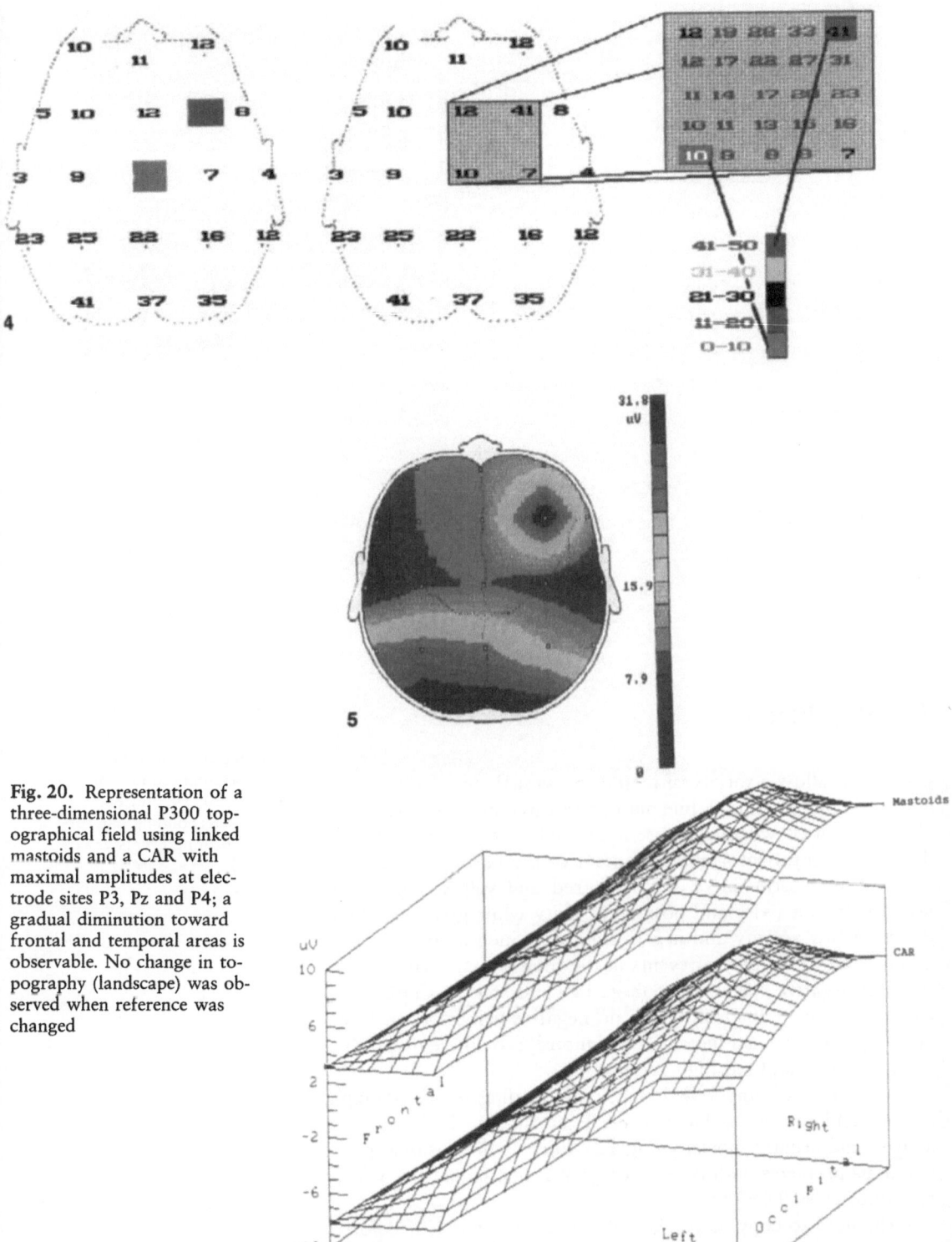

Fig. 20. Representation of a three-dimensional P300 topographical field using linked mastoids and a CAR with maximal amplitudes at electrode sites P3, Pz and P4; a gradual diminution toward frontal and temporal areas is observable. No change in topography (landscape) was observed when reference was changed

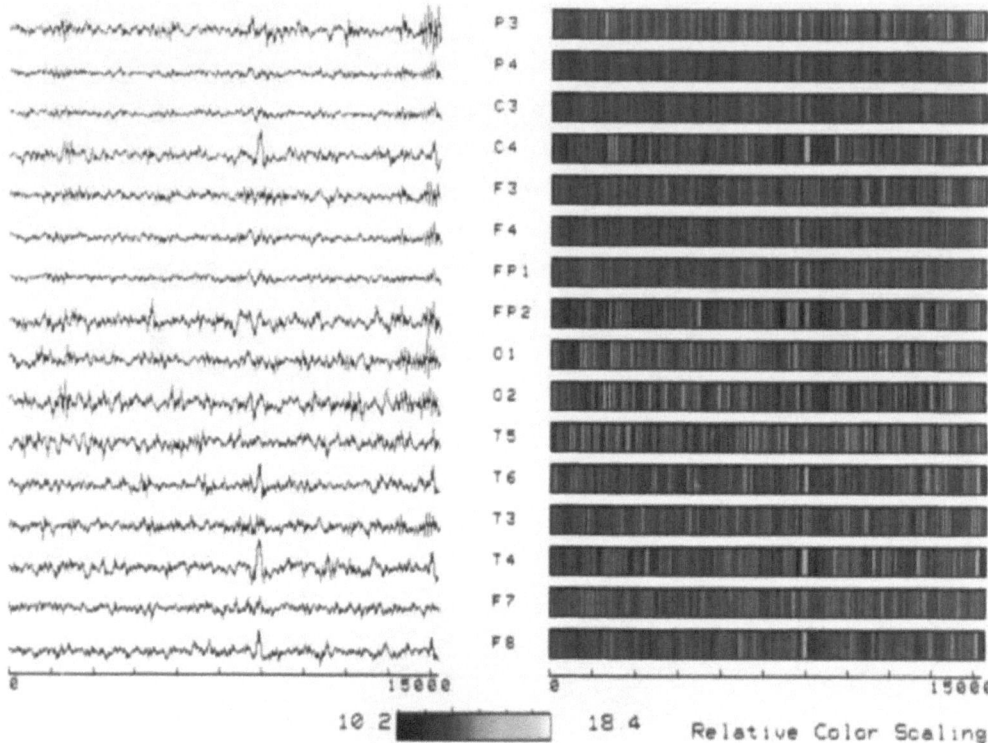

5.6 Map Features

Computers allow a variety of map displays such as colored maps, gray-scale maps, contour-line maps, and three-dimensional maps. The question of whether to use a colored or a gray-scaled map depends upon the information content of the particular map. By convention (Herrmann et al. 1989), red and yellow colors are used for high activity and positive polarity while green and blue represent low activity and negative activity. Such maps are commonly used for display of events in the time domain where positive and negative values occur (e. g., EPs and spikes). Calibration bars in frequency maps in which no negative values occur, however, begin at zero and increase continuously; for these maps blue represents low and red high activity.

In the case of gray-scaled maps dark shading represents high activity and positive polarity while light shading indicates low activity and negative polarity (Fig. 22). In three dimensional presentations problems with colored or gray-scaled maps and polarity do not exist (Dierks et al. 1989 a).

The need to copy maps also influences the choice of colored or gray-scaled maps, for copying a black-and-white display is easier and less expensive (Buchsbaum et al. 1982 c).

Fig. 21. Line and bar format displays. *Left,* 16 channel EEG recording with paroxysmal events pronounced in channels C4, Fp2, T6, T4, and F8. *Right,* corresponding bar formats, indicating events as yellow bars

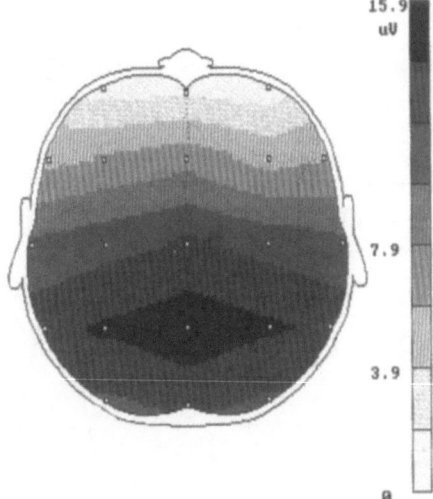

Fig. 22. P300 illustration using gray scaling. For representation of amplitudes eight different steps were produced by increasing the number of black spots per unit

Each map should be accompanied by a calibration bar indicating (a) amplitude ranges (µV) in the case of EPs and EEG events in the time domain, (b) activity ranges (µV) in the case of EEG frequencies in the frequency domain and (c) ranges for statistical values from z statistics (standard deviations), t-tests (t values), and significance levels (p values). In the case of colored maps 16–32 color steps are generally used. With 16 steps and a calibration bar that indicates 16 µV as maximal value, each step represents a range of 1 µV. Depending on the activity the calibration bar can be adapted according to the altitude of the signal.

5.7 Mapping of Evoked Potentials

In the case of EP, positive and negative amplitude values are usually mapped by averaging single trials of EPs. Because the amplitude relationship between channels is always the same, and the absolute amplitude values in each channel depend upon choice of reference, it is advisable to use reference-free methods for evaluation of latencies, amplitudes, and topography. An example of this is the spatial standard deviation (global field power, GFP; Lehmann and Skrandies 1980).

5.7.1 Latency and Amplitude Determination for EPs and ERPs

With EPs of early latencies with-well defined peaks (e. g., early auditory EPs) maxima and minima of the peaks can be used for

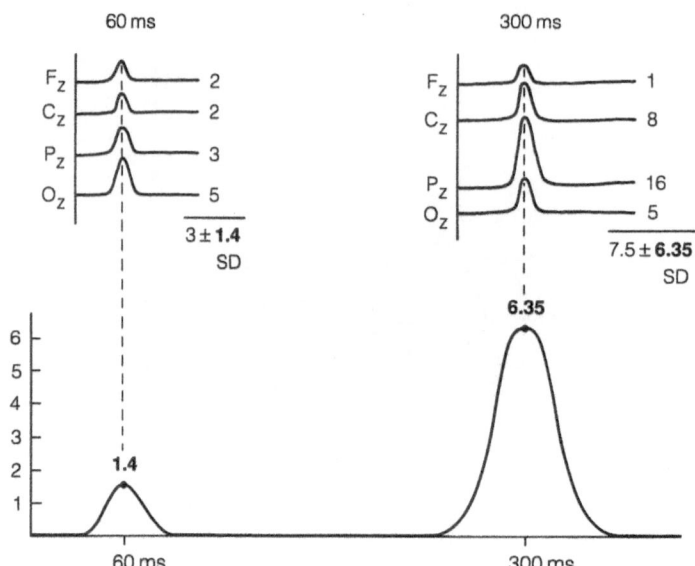

Fig. 23. Schematic drawing for explanation of GFP. For simplification only four channels have been used. At a latency of 60 ms a mean of 3.0 µV and standard deviation of 1.4 µV were calculated and at 300 ms a mean of 7.5 µV and standard deviation of 6.35 µV

latency determination. For early auditory EPs only a few electrodes are necessary due to the fact that "far-field potentials" are recorded. A display of spatial distribution is at present not meaningful for clinical purposes.

When recording EP and ERP (event-related potentials) series with late latencies at multiple locations on the scalp a large number of possible waveforms results. With 20 channels maximally 380 waveforms [n(n-1)] may result, and an additional 40 (2n) can be calculated using CAR. What is the correct waveform for determining latency, amplitude, and topography? Ideal would be reference-free measurements. Lehmann and Skrandies (1980) have therefore proposed the GFP, mentioned above, as a way to measure latency in a reference-free manner and to obtain a reduction in spatial data. GFP is the spatial standard deviation at each recorded point in time and thus a measurement of the field variability in space (Fig. 23). GFP is a reference-independent measurement since relative differences between electrodes at a defined point in time do not change regardless of which reference is used; this is due to the fact that the electrical landscape is produced by the brain and is not influenced by electrode locations.

Determining the strength of response times of maximal GFP in the EP series establishes the latencies of the components (Fig. 23). Component strength (amplitude) is, i. e. the GFP or the potential difference between the maximal and minimal value in the field at the time that the GFP reaches its maximum.

6 Storing of Data

Magnetic or optical media can be used to store EEG signals. Hard-disk systems are available that can store up to 300 Mbytes. Table 4 presents an overview of storage capabilities.

Small inexpensive magnetic disks (diskettes) are widely used for personal computers, but only a small amount of EEG data can be recorded on each disk. These disks are better used for storing a compressed representation of data such as brief epochs of EEG or frequency and amplitude maps. Other storage media are videocassettes. Routine VHS or Betamax type videotapes can be used. An advantage is the fact that patients can be filmed simultaneously with the EEG (e. g., via an infrared camera for documentation of one night's sleep). A disadvantage is the slow access time. An improved method is the storage of digitized EEG or EP data on digital audio tapes.

In most countries it is required that EEG data be stored for 10 years. This can be done as a paper EEG, as microfilm or in a digitized manner. Magnetic disks eventually become demagnetisized and loose the stored data after some years. Optical disks should theoretically store EEG data forever. When using Winchester drives (non changable hard disks; storage from 10 to around 300 Mbytes), back-up of data must generally be done on streamer (magnetic tape). Streamer, however, has the great disadvantage of very slow access time for data. Recommended at present are 1.2 Mbyte disks, 44 Mbyte changeable hard disks, and optical drives when purchasing a commercial system for mapping EEG data.

Table 4. Storage capacities of magnetic and optical disks and their approximate storing times for EEGs depending upon sampling rate and number of recording channels

360	Kbyte disk	2 min 30 s
720	Kbyte disk	5 min
1.2	Mbyte disk	8 min
1.44	Mbyte disk	10 min
20	Mbyte hard disk	2 h
44	Mbyte disk changeable hard disk	4 h 30 min
200	Mbyte optical disk	22 h

7 Statistical Procedures

EEG data rarely follow Gaussian distribution; therefore it is advisible to transform data before statistical treatment to achieve gaussianity, i.e. by logarithmic transformation (Gasser et al. 1982).

Appropriate evaluation of map alterations and their significance is important for data on single patients and for group results. In comparing a series of patients with a normal group, it is essential to have a database on a homogeneous group. For the evaluation of a single case the z statistic can be used (Fig. 24). This is calculated on the basis of deviation of a patients test value from the control group mean and is expressed in units of standard deviations:

$$z = \frac{xg-xi}{sd}$$

where xg is the arithmetic average of the group, xi is the value of the single case, and sd is standard deviation of the group. The z statistic offers a simple method for evaluation of a single case. However, it also has disadvantages, and significances may occur where none exist.

Comparisons of a single case to group data by such means as the z statistic make use of a database that contain information on the patients in terms of their age, sex, handedness, diseases, and type and dosage of medication.

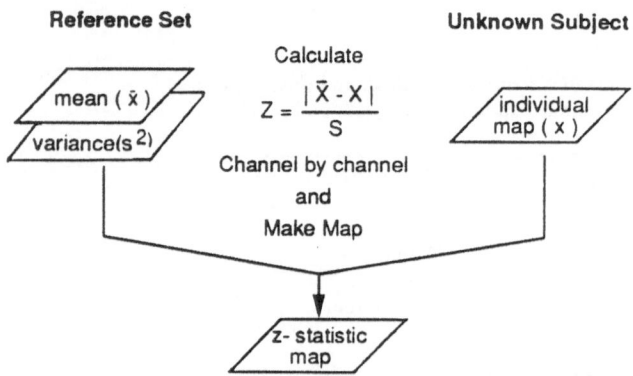

Reference Set   Calculate   **Unknown Subject**

mean ($\bar{x}$)

variance(s^2)

$$Z = \frac{|\bar{X} - X|}{S}$$

individual map (x)

Channel by channel
and
Make Map

z- statistic map

Fig. 24. Calculation of a z statistic map. The z statistic represents the extent to which (in terms of SD) an individual observation differs from the mean of a reference set. (According to Duffy et al. 1989)

Fig. 25. Calculation of a t statistic map. Student's *t* statistic quantifies extent of difference between two sets of measures, taking into account the difference between group means and the variance within each group. (According to Duffy et al. 1979)

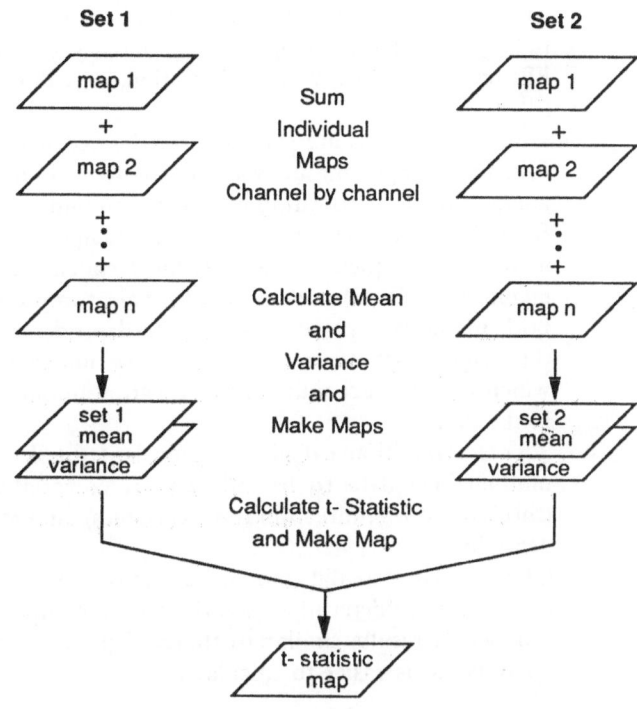

A sophisticated method for the classification of single cases has been developed by John et al. (1977) under the term "neurometrics"; this is discussed further in Chap. 10.

Prior to comparison of a patient's test values to group data, a data reduction is recommended, for example, by dividing the EEG into different frequency bands (delta, theta, alpha, and beta) and by procedures described below.

Statistical procedures used for group comparisons can be divided into exploratory/descriptive and confirmatory ones.

Exploratory/descriptive methods include:
- Student's *t* test (significance probability mapping; Duffy et al. 1981) in which the pixels are presented as *t* or *p* values (Fig. 25). This test is available in most commercially available mapping systems (Taira et al. 1986).
- All statistical methods with multiple testing without Bonferroni correction (e. g., Wilcoxon test, linear regression)

Maps using z, *t*, or *p* statistics cannot be used for confirmatory purposes in demonstrating an effect.

Confirmatory methods include:
- Analysis of variance (ANOVA) with one or more factors
- Multivariate analysis of variance (MANOVA) and discriminant analysis

—   *t* test, Wilcoxon test, or Mann-Whitney U test after substan-
    tial data reduction (e. g., GFP for latency determination in
    EPs or correction for multiple testing, e. g., Bonferroni).

A number of methods are available for data reduction:
—   Principal component analysis and the recently developed sin-
    gular value decomposition reduce the amount of redundancy
    from multichannel data. Principal component analysis is
    used to set frequency ranges within frequency spectra (Her-
    mann et al. 1989) and to select single components of EPs.
    Both methods explore data for interdependence and reduce
    the original set of variables to a few independent factors
    which together account for the greatest amount of variance
    in the data.
—   Source estimation calculates equivalent dipoles and allows
    multichannel data to be reduced to six variables: location
    (three variables), direction (two variables), and strength (one
    variable).
—   GFP calculates spatial standard deviation (see Sect. 5.7.1) and
    can be used to determine essential spatial components in EPs.
    This yields results similar to those of principal components
    analysis but is easier to calculate.

It is important to differentiate between exploratory, descriptive,
and confirmatory analyses. The way in which an analysis has
been performed can be indicated by an index (p_e, explorative; p_d,
descriptive; p_c, confirmative) according to recommendations of
the Mapping Committee of the German EEG Society (Hermann
et al. 1989). Statistical calculations should always be performed
using values measured at electrode points and not by means of
interpolated pixels. Statistical methods for multichannel EEG
and EPs is a controversial issue, and no ideal method has yet
been found. Advanced procedures are shortly mentioned in
Chap. 10.

8 Practical Application:
Findings in Normal Subjects

8.1 Introduction

Mapping of the electrical activity of the brain may be divided in two main categories: analysis in the time domain and analysis in the frequency domain. A difference in the case of mapping as compared to conventional EEG/EP recordings is the fact that data of both these domains are transferred to the spatial domain by measuring the amount of activity over a restricted number of electrode sites and using interpolative methods for interelectrode points (pixels) to produce a continuous visual image. In the time domain map features of raw EEG/EP data are presented, as shown in Table 2. In the frequency domain, a computer algorithm (as described in Sect. 5.4) converts EEG data to a series of frequency components and their associated amplitudes.

8.2 EEG Features in the Time Domain

The simplest procedure is the topographical display of the alpha rhythm with its phase reversal which has already been described by Berger (1933; Fig. 26). Mapping of background activity in the time domain is of minor value in the clinical application compared to frequency mapping. More important is the topography of special events such as spikes, sharp waves, focal slow wave activity, and features correlated to sleep. Examples can be found in the following sections.

8.2.1 Dipole Estimation

A dipole is a simplified equivalent of simultaneous discharges of multiple neurons. Proper localization of potential sources for dipoles is important (Scherg and von Cramon 1985). In conventional EEG recordings a phase reversal in the bipolar setting may indicate the precise location of a focus. In mapping, dipoles may facilitate the search for sources. If there are no dipoles the source

a

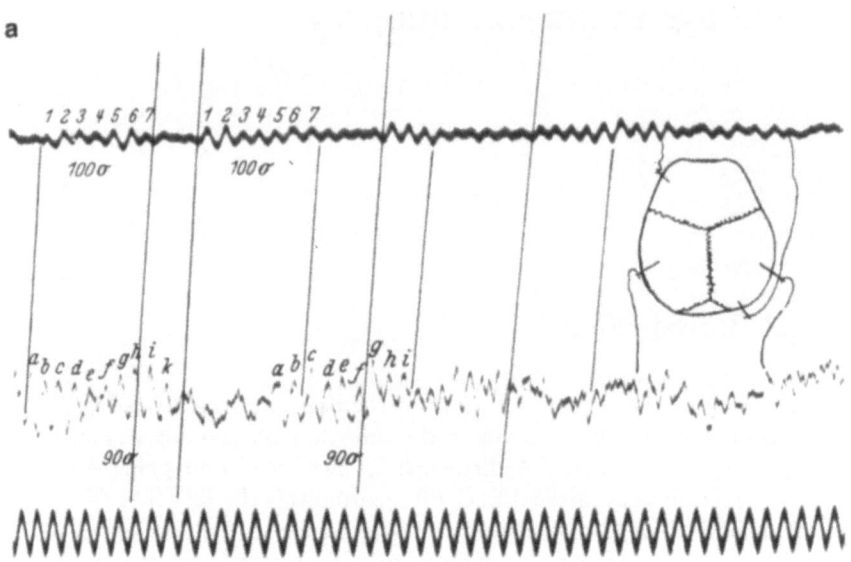

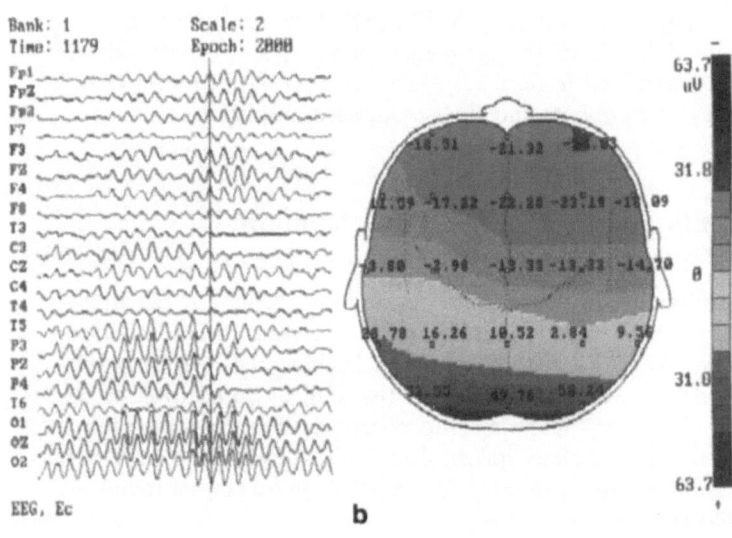

Fig. 26. a Original EEG traces from H. Berger recorded from the left forehead/occiput (upper EEG) and from both vertex areas (from Berger 1933). **b, c** Topographic display of alpha waves in the time domain. If the cursor is fixed upon a positive peak occipitally negative values appear frontally according to a phase reversal of 180°. **d** As one moves gradually along the time axis, a counterclockwise movement of positive and negative fields is observed with a mere positive field occipitally between 539 and 546 ms

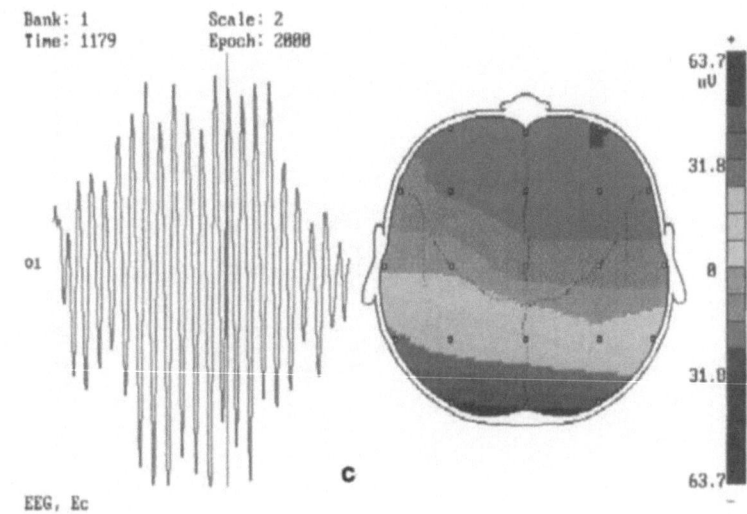

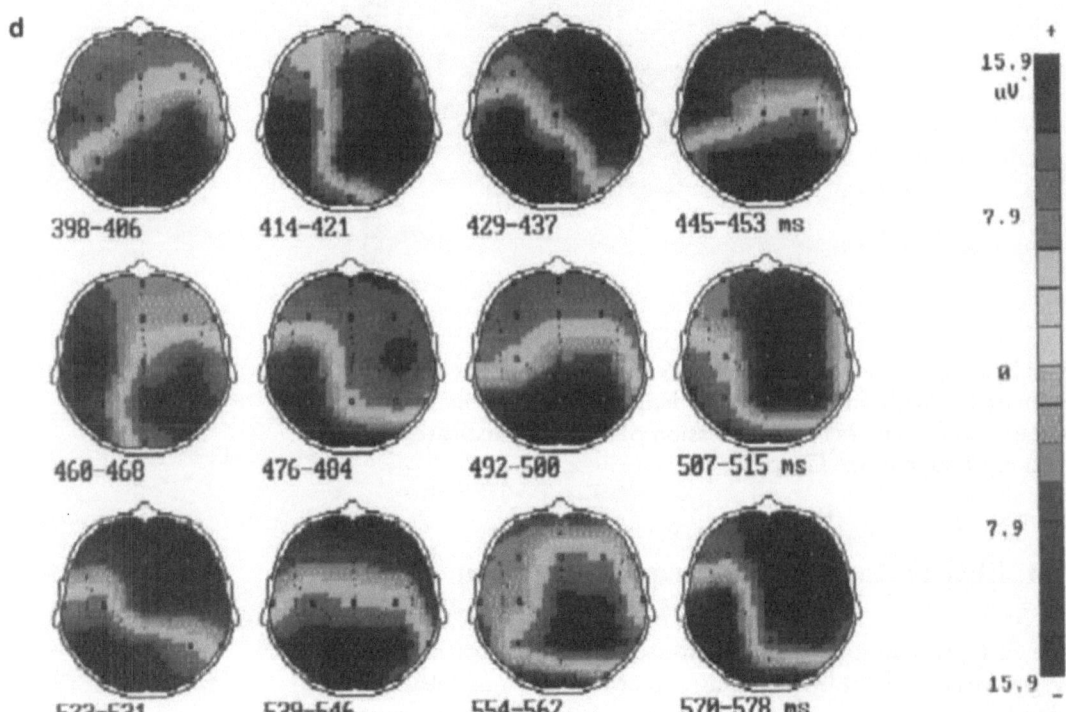

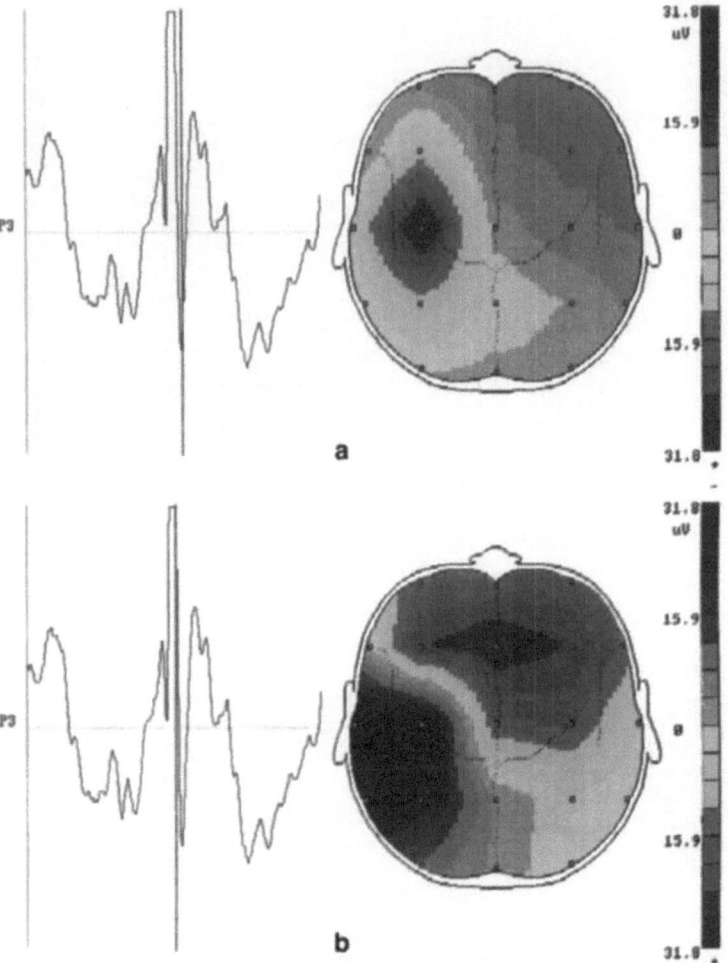

Fig. 27. a Spike presentation with a positive maximum of epileptic activity at C3. Due to the presence of only on positive field a radial source of activity may be assumed directly under electrode C3. b Same spike presentation with indication of a single tangential dipole with a negative maximum around T3, C3 and T5 and a positive one around Fz, assuming the source to be located between the two sites

may be directly under the electrode; otherwise, the source can be estimated visually in terms of the length of the dipole and the voltage values (Fig. 27). For discussion of more sophisticated estimation of sources see Chap. 10.

8.3 EEG Features in the Frequency Domain

The FFT procedure is described in Sect. 5.4. The most common target variables for EEG mapping are power (μV^2) or mean amplitudes (μV) in four to six frequency bands. The frequency range for analysis extends from 0.5 to 32 Hz or higher, depending on sampling rate. The main frequency ranges are as follows: delta, 0.5–3.5 Hz; theta, 4.0–7.5 Hz; alpha, 8.0–11.5 Hz; beta, above 12.0 to the Nyquist frequency. The Mapping Committee

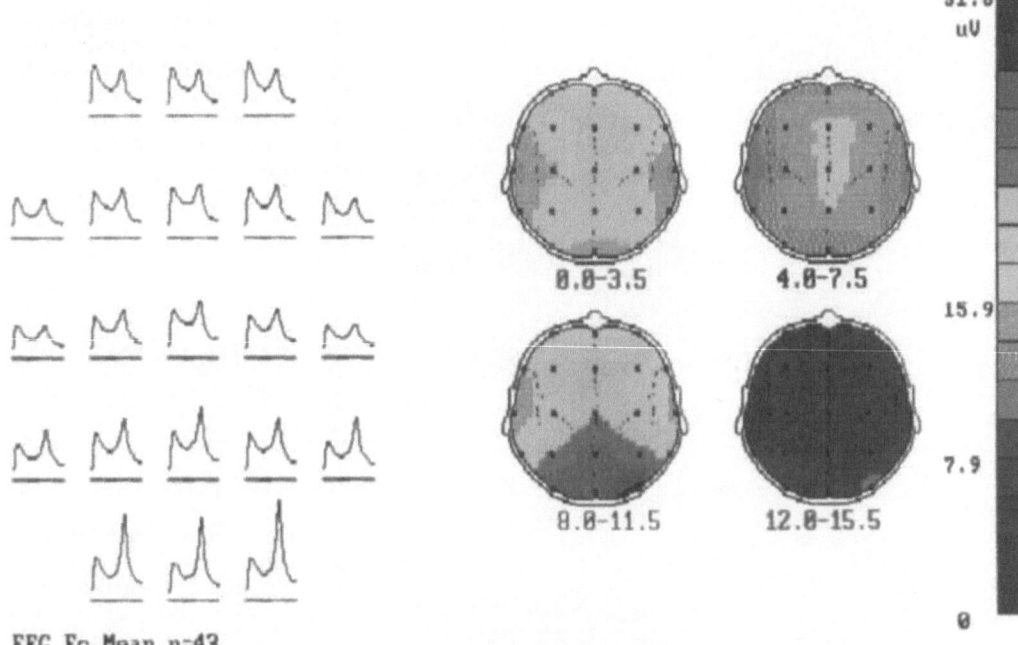

Fig. 28. Topographic display of the four main frequency bands (delta, theta, alpha, beta) in a sample of 43 normal subjects. *Left,* frequency spectra with evident alpha peaks especially at occipital and parietal sites. *Right,* corresponding maps

of the German EEG Society (Herman et al. 1989) recommends that at least four to six frequency bands be represented (delta, theta, alpha 1, alpha 2, beta 1, beta 2, and beta 3). The clinician should be able to inspect activity in frequency steps of at least 0.5 Hz. Figure 28 shows frequency maps in the main frequency domains.

8.4 EP Features

Methodological procedures for measuring EPs are beyond the scope of this atlas. For these the reader may refer to Maurer et al. (1982, 1988 e and 1989 c) and Lowitzsch et al. (1983).

8.4.1 Mapping of Visual Evoked Potentials

Visual EPs and their topographical fields are of singular importance in the clinical context (Bourne et al. 1971; Allison et al. 1977; Halliday et al. 1977; Skrandies and Lehmann 1982; Adachi-Usami and Lehmann 1983; Edwards and Drasdo 1987). A visual stimulation can be provided by a change in brightness (flash) or a change in contrast (pattern shift). Single flashes from a

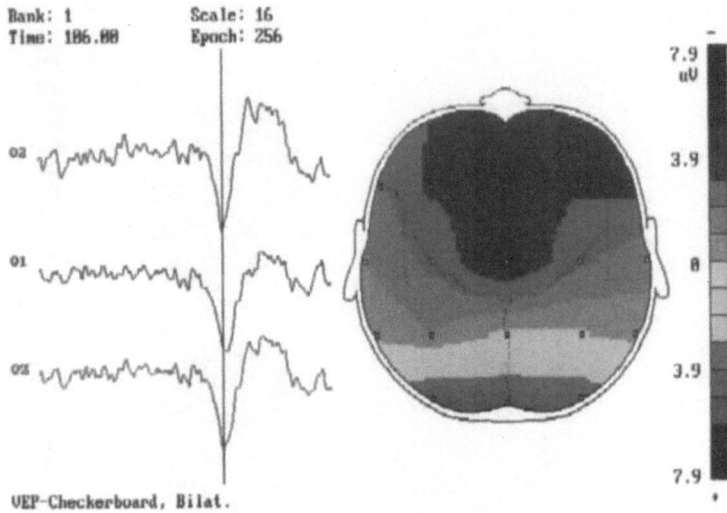

Bank: 1   Scale: 16
Time: 106.00   Epoch: 256

O2

O1

Oz

UEP-Checkerboard, Bilat.

7.9 uV

3.9

0

3.9

7.9

Fig. 29. Map of visual EPs. *Left,* at a latency of 106 ms maximal positive peaks were reached at recording sites O1, Oz, and O2; *right,* corresponding topographical map calculated from all 20 electrode sites

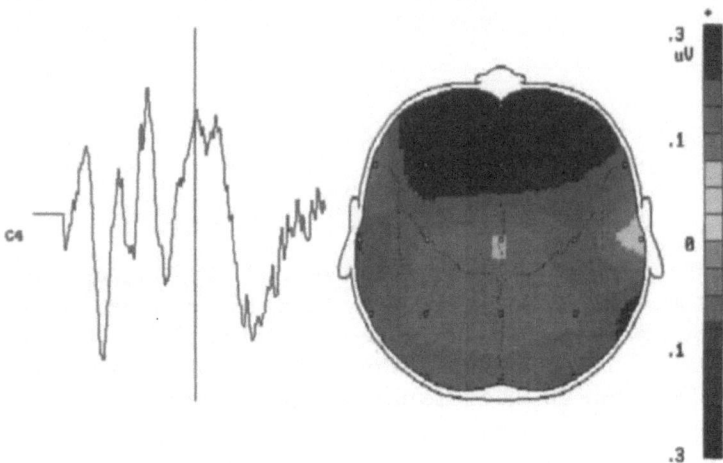

C4

.3 uV

.1

0

.1

.3

Fig. 30. Topographical display of wave IV of early auditory EPs. *Left,* single trace at C4; *right,* corresponding map with 20 electrodes. In this subject a maximum of positivity (300 nV) appeared at frontal sites

strobe lamp are triggered at a frequency of 1–2 cps. The distance between the eye and the bulb is approximately 20–30 cm.

The contrast stimulation is produced by a change in pattern (e. g., rotating a checkerboard or a bar graph). Figure 29 shows visual EPs and their topography. A stimulus design developed by Dierks and Maurer (1989) allows the elicitation of a visual evoked P300.

8.4.2 Mapping of Auditory Evoked Potentials

Approximately 30 components can be elicited and mapped after acoustic stimulation (Scherg and von Cramon 1985; Grandori 1986; Comacchio et al. 1988; Deiber et al. 1988; Jacobsen and Grayson 1988; Karniski et al. 1989). Mapping of early compo-

Fig. 31. a Topographical display of the N100 (N1) component (92 ms). *Left,* a single trace at Cz; *right,* corresponding map. Maximum negativity appeared at electrode site Cz (common average reference). b Topographical display of the P180 (P2) component (180 ms) with a single trace at Cz on the left side and a corresponding map on the right side. The maximum of positivity appeared at electrode site Cz (common average reference). c Topographical display of the P300 (P3) component (320 ms). *Left,* a single trace at Cz; *right,* corresponding map. Maximum positivity appeared at electrode site Pz (common average reference)

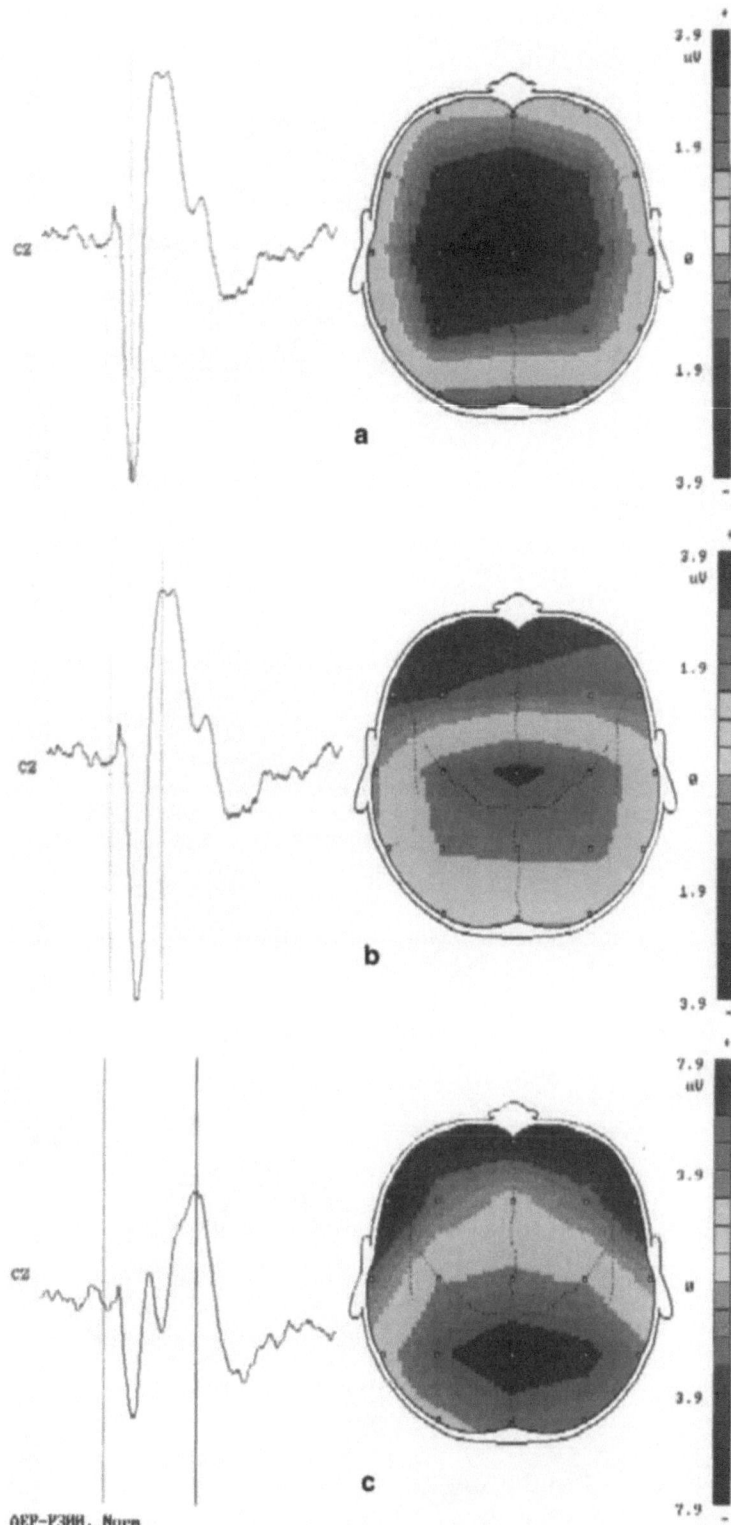

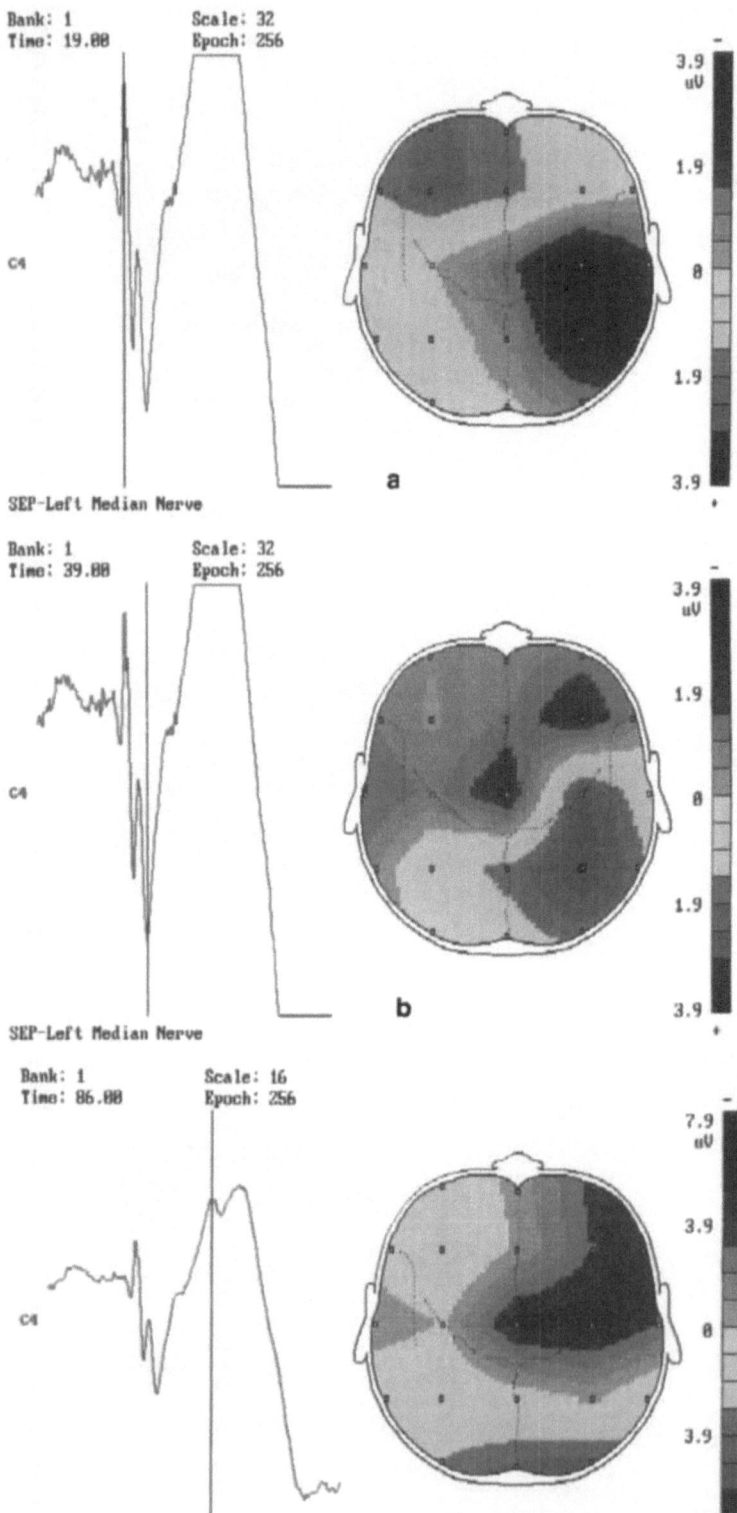

Fig. 32. a Topographic display of N20 SEP component. *Left*, single trace at C4; *right*, corresponding map with maximum negativity around C4 and P4 electrode sites. b Topographic display of P40 SEP component. *Left*, single trace at C4; *right*, corresponding map with maximum positivity around P4 electrode site. c Topographic display of P80 SEP component. *Left*, single trace at C4; *right*, corresponding map with a maximum of negativity around Cz and C4 electrode sites

Fig. 33. Topographic display of a contingent negative variation recorded with 20 channels

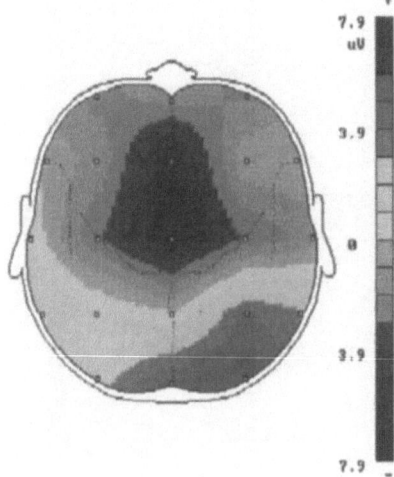

nents (e. g., early auditory EPs) is at present of minor clinical value because of the "far-field" character of early auditory EPs and their widespread field upon the head-without clinically relevant topographical features (Fig. 30). Of interest is the mapping of exogenous (N1, P1, N2, P2) and endogenous components (P300, slow wave). Corresponding EPs and their topography are presented in Fig. 31 a to c.

8.4.3 Mapping of Somatosensory Evoked Potentials

Recording of somatosensory EPs offers a functional test of the integrity of the specific lemniscal somatosensory system. The primary cortex is organized somatotopically, which is important for mapping (Goff et al. 1977, Duff 1980 b; Kakigi and Shibasakin 1983; Jones and Power 1984; Desmedt and Bourguet 1985; Tsuji and Murai 1986; Desmedt et al. 1987; Desmedt and Tomber 1989). The topography of main exogenous components is shown in Fig. 32 a to c).

8.4.4 Mapping of Contingent Negative Variation (CNV) and in Response to Olfactory and Chemosensory Stimulation

The stimulus design for provoking CNV has been thoroughly studied by Zappoli (1988). CNV can also be studied by topographical display (Fig. 33). Topographical studies and maps produced by olfactory and chemosensory stimulation have been published by von Toller and Reed (1989).

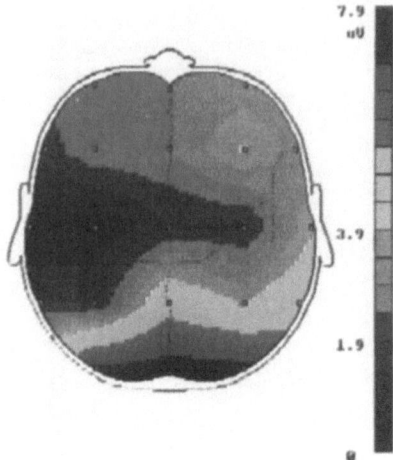

Fig. 34. Example of an EEG map after motor activation due to movements of the right fist. Due to motor action an attenuation of alpha at 10 Hz with a maximum at T3 and C3 occurred

8.5 EEG Mapping After Sensory, Motor, and Mental Activation and due to Psychotherapeutic Interventions

The activated or dynamic EEG (Pfurtscheller 1986) is suitable for investigating perceptual, motor, and cognitive cerebral functions and may be used in behavioral studies (Freeman and Maurer 1989 a). It has long been known that after movements such as the opening and closing of the hand contralateral desyncronization or alpha blocking occurs, which is frequently accompanied by waves in the beta range. In psychiatric disorders it is advisable to select activation methods related to the psychopathological findings. Figure 34 shows characteristic EEG changes for different activation methods. The short analysis time in mapping is an advantage in removing problems of maintaining a cognitive steady state that may arise over a relatively long measuring period (autogenic training: Bostem and Degossely 1978; Ormejohnson and Gelderloos 1988; Dierks et al. 1989 b; music: Breitling et al. 1987; words: Brown and Lehmann 1979).

8.6 Sleep Features

The topographical display of reduction in vigilance and the spatial display of sleep features are new areas in mapping (Findji et al. 1981; Buchsbaum et al. 1982 b; Maurer et al. 1989 a). Figure 35 presents examples in the form of topographical hypnograms. Other examples are shown in Figs. 36 and 37).

Fig. 35 a–e. Topographic displays of hypnograms at different sleep stages which were evaluated according to rules of Rechtschaffen and Kales (1968). The 8 head formats demonstrate frequency features between 0 and 20 Hz. **a** In the awake state a typical alpha field was observed between 7.5 and 9.5 Hz. **b** During stage I an attenuation of alpha and a slight increase of delta took place. **c** During stage II alpha was no longer present. Instead, delta and theta increased. Note the sigma spindles between 12.5 and 14.5 Hz with a maximum at Pz. **d** During stages III and IV (a differentiation between the two stages was topographically not possible) slow delta predominated with a maximum at Fz. **e** During REM stage generally attenuated activity in all frequency bands with evident slow activity due to rapid eye movements was observed

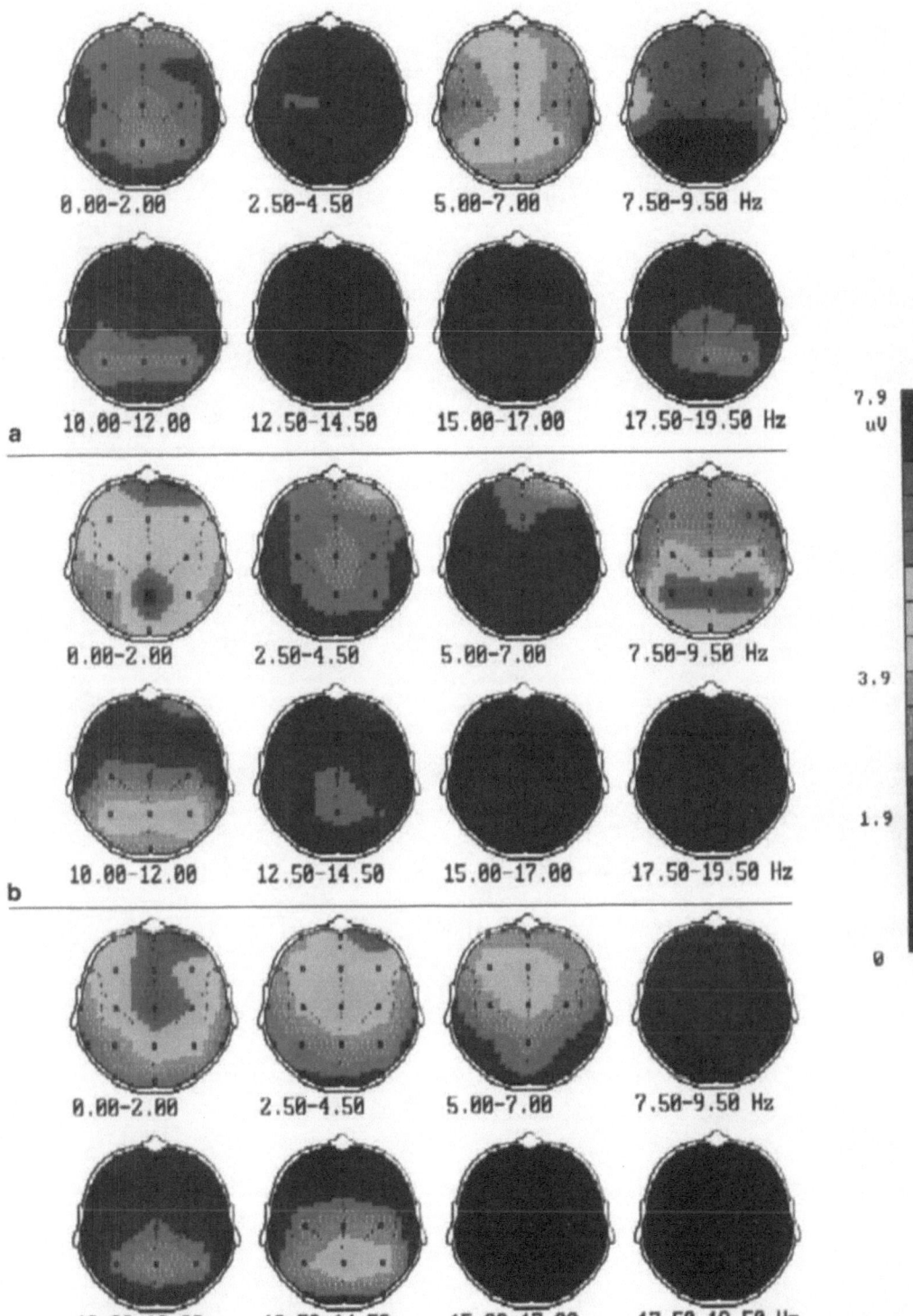

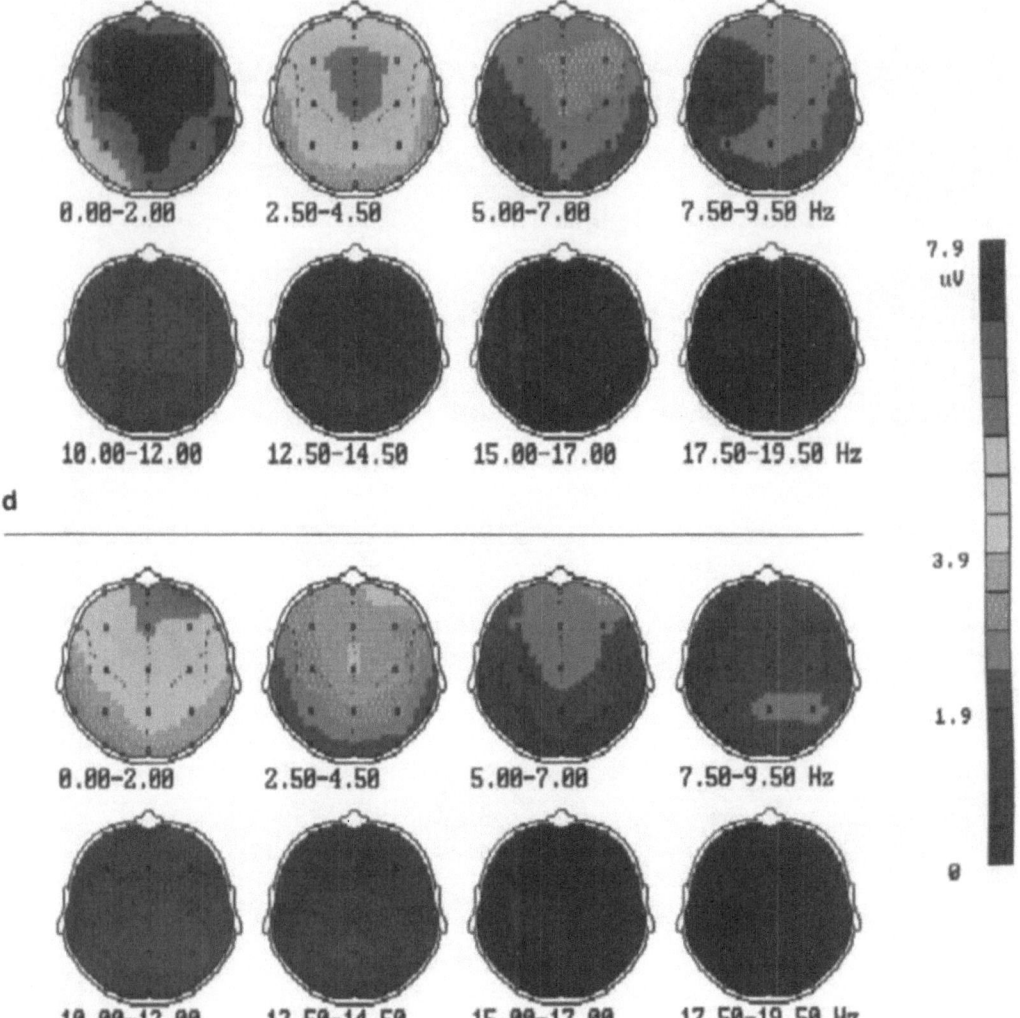

Fig. 36. Topographic display (time domain) of a vertex wave with maximal negativity at cerebral point Cz

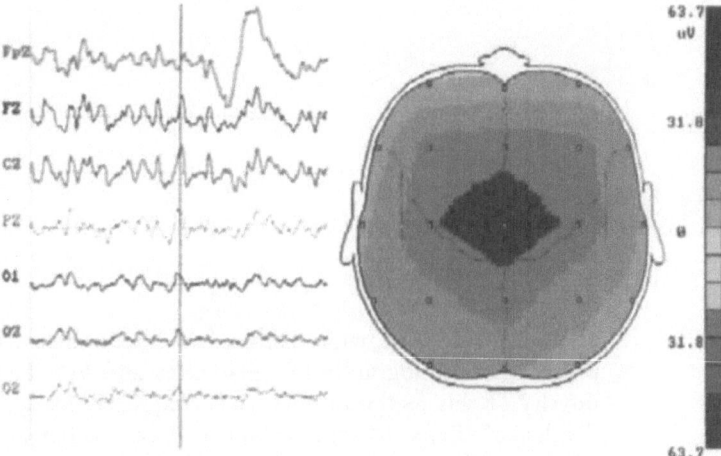

Fig. 37. Topographic display (time domain) of a K complex with a vast negative field around cerebral points Cz and Pz

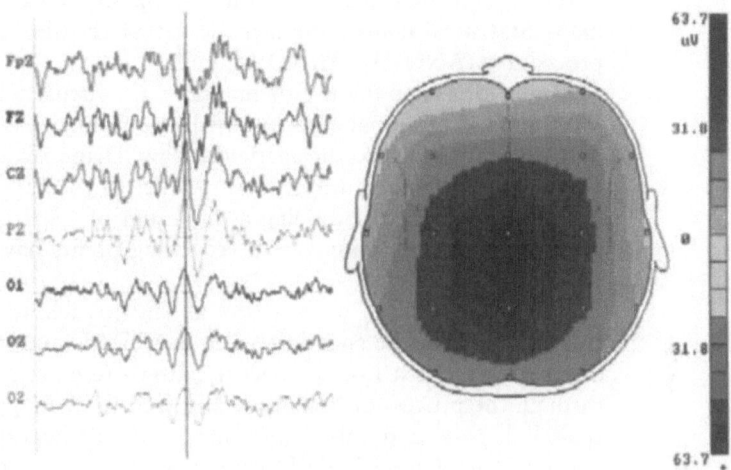

9 Findings in Diseases

9.1 Introduction

The maps below showing diseases have been chosen to show the pathological topography of EEG states and EPs. Because of the novelty of this method reference is frequently made below to individual findings. To some extent, however, it has also been possible to study groups of patients with defined diagnoses. In these cases group means of controls as well patient data are shown. Statistical differences were determined by t tests for exploratory purposes and imaged in the form of topographical significance maps. Statistical confirmation made use of the above-mentioned procedures (ANOVA, MANOVA, etc.).

Indications for the use of mapping are virtually the same as those for conventional EEG but with the distinct advantage that data are transferred to the spatial domain. Diagnosis with EPs involves a number of entirely new points, since mapping can be performed only after applying a minimum of 16 electrodes. Until now only a small number of recording points have been used for EPs.

It is impossible to present here all the possible topographical features that may be encountered in psychiatric and neurological diseases. There is a loss of specific pattern (e. g., shape of spike) through the process of amplitude and frequency mapping and its spatial display, and this results in a nonspecific topographical feature. The description of diseases in neurology is therefore dominated by the topography of local frequency, and amplitude differences and by paroxysmal events. Psychiatric disease entities such as psychoses and dementia are described below. A frequent use is the mapping of EEG and EPs after the administration of drugs.

9.2 Evaluation of EEG and EP Maps

Before considering the clinical application, this section demonstrates how maps can be evaluated in a way effective for the purposes of psychiatrists, neurologists, neuropsychologists and physicians who may be involved by mapping.

The difference of mapping to conventional EEG and quantitative EEG is that data are displayed in the spatial domain by giving each of the 19 or more electrodes of the 10–20 system a position on the map. The amount of activity under each electrode is then used to interpolate values for the interelectrode points (pixels). This limited number of actually measured values and thousands of interpolated pixels produce a topographical map of electrical activity. Mapping is useful only if it provides information which cannot be gained by inspection of a conventional EEG or EP recording or by application of quantitative EEG.

How should EEG or EP maps be considered for clinical evaluation? Conventional EEG uses uni- and bipolar recordings for evaluation. In the unipolar state (e. g., Goldmann reference) amplitudes and frequencies are determined by visual inspection, and topographical evaluation depends more or less upon the spatial imagination of the interpreter in merely looking at data from 8–16 channels. Bipolar recordings (e. g., transverse and longitudinal electrode arrays) were carried out to determine, for example, phase reversals for proper focal estimation. With mapping the electrical activity is measured either under linked ear or mastoid (A1/A2) conditions or with a single cephalic electrode (e. g., Cz). In addition, the calculation of reference-independent values is recommended, which enables a standardized comparison of maps.

For interpretation of amplitude and frequency maps there are several conditions which should be kept in mind. *One* is that, with the interpolation method (linear interpolation) there are only few electrode sites on the head but thousands of calculated pixels, meaning that maximal or minimal activity is always at electrode sites. *Second,* if a frequency change or an amplitude transient occurs between two or four electrodes, the event may be missed in the case of long interelectrode distances (Fig. 5), or a similar amount of activity may occur at surrounding electrodes, giving an impression of a large electrical field. The activity shown in a map in such a case is always lower than the real signal strength. This disadvantage of inaccurate measurement and localization of activity can be resolved only by applying more electrodes to the head, with shorter interelectrode distances, thus reducing spatial aliasing. *Third,* frequency maps are dependent upon the reference (see Sect. 5.4). The interpreter may therefore be confronted with numerous frequency maps since frequency spectra recorded with different references cannot be compared with each other. It is therefore advisable for the user to maintain an established reference for recordings. *Fourth,* the calculation of CAR, in the case of mapping of local transients such as spikes and EPs, emphasizes the dipole character of the electrical field. This may be helpful for visual source localization. Figure 42, i. e., shows a dipole with a positive and a negative maximum with similar magnitudes. A tangential equivalent dipole between the two electrodes may be assumed in this case, while radial ones tend to produce single positive or negative fields. And, *finally,* the color of maps may mislead clinicians to assume a pathological condition where none in fact exists. Especially the appearance of bright red colors psychologically lends significance. In *summary,* map evaluation should not concentrate solely on local maxima and minima of voltages or on activity and ignore map features of the whole head with its underlying dipoles.

9.3 Clinical Examples

9.3.1 Introduction

As noted above, disease findings by mapping cannot be described exhaustive merely by considering each entity; such an approach is common in textbooks on conventional EEG findings. Due to the transformation of data from the time to the frequency and spatial domains the loss of specificity of EEG pattern is accompanied by a multitude of nonspecific spatial data of EEG and EP features. It is therefore necessary to describe spatial peculiarities of local frequency and amplitude differences, EEG transients, and abnormal field distributions of exogenous and endogenous EPs. This is done below in single case presentations in diseases where individual topographical features are to be expected and in demonstrations of group data in disease entities with a defined lesion pattern.

9.3.2 EEG Mapping of Local Frequency and Amplitude Differences

Local frequency and amplitude differences occur in clinical conditions such as brain tumors, vascular processes, inflammatory disorders, and traumatical injuries.

9.3.2.1 States Causing Increased Intracranial Pressure (Brain Tumors)

Brain mapping opens new possibilities for functional diagnosis of states that cause increased intracranial pressure (Nagata et al. 1985). This atlas presents results in a series of patients with intracranial tumors. Table 5 presents the diagnoses and results gained by CT and EEG mapping. Figure 38 shows a representative case with a simultaneous display of CT and EEG maps. For statistically reasons Z-statistics were used.

Table 5. Demographics, diagnoses, and results of CT and EEG mapping in 14 neurosurgical patients

Case	Age (years)	Sex	Diagnosis	Brain sectors (see Scheme below)						
				1	2	3	4	5	6	
1.	25	M	Right frontoprecentral astrocytoma	+	+					EEG
				+	+					CT
2.	26	M	Left temporal hemangioma	+				+	+	EEG
								+	+	CT
3.	21	M	Epileptic seizures	+	+	+			+	EEG
										CT
4.	47	M	Bifrontal astrocytoma	+	~			+	+	EEG
				+				+	+	CT
5.	33	F	Right occipital ependymoma, left parietal tumor		+	+				EEG
					+	+	+			CT
6.	56	F	Right occipital metastasis		+	+				EEG
					+					CT
7.	23	F	Right epileptic focus	+	+	+			+	EEG
										CT
8.	57	F	Right frontotemporal glioblastoma	+	+					EEG
				+	+					CT
9.	64	F	Right frontal tumor	+	+					EEG
				+	+					CT
10.	48	F	Right occipital hemangioma	+	+	+				EEG
					+	+				CT
11.	22	F	Right cerebellar astrocytoma			+				EEG
						+	+			CT
12.	62	F	Right frontal falx meningioma	+	+				+	EEG
				+					+	CT
13.	64	F	Left frontolateral glioblastoma	+				+	+	EEG
				+				+	+	CT
14.	30	M	Left precentral astrocytoma					+	+	EEG
								+	+	CT

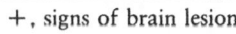

+, signs of brain lesion

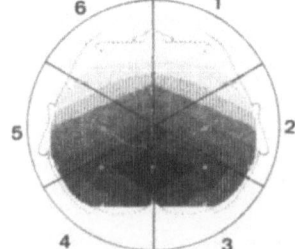

Head format with the six brain sectors used in the study to achieve a statistical comparison between CT and EEG findings. Numbers correspond to the six brain sectors in Table 5.

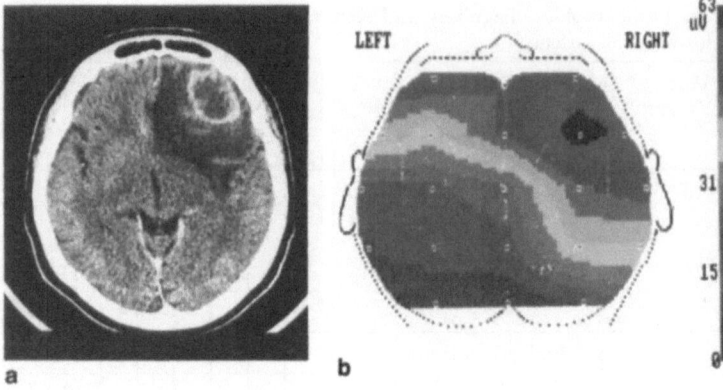

Fig. 38 a, b. EEG map (b) and CT (a) in a patient with a right frontotemporal glioblastoma (case 8 in Table 5). EEG map showed slow activity in the delta band (2 Hz) especially at F4 and C4

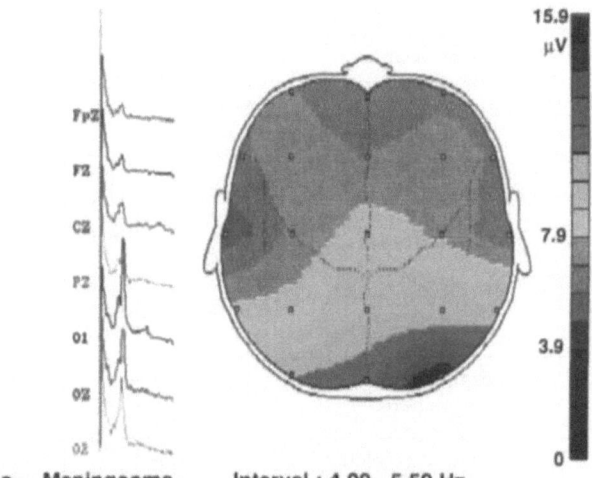

a   Meningeoma   Interval : 4.00 - 5.50 Hz

Fig. 39 a, b. EEG maps in a patient suffering from a right sided occipital meningeoma (from Hamburger 1989).
a Increased slow activity (4.0–5.5 Hz) at the site where a meningeoma was later found, with theta activity pronounced at electrode O2.
b *Left,* before removal of meningeoma, right-sided attenuation of alpha activity (alpha reduction); *right,* after removal, restitution of a symmetrical alpha field on both sides

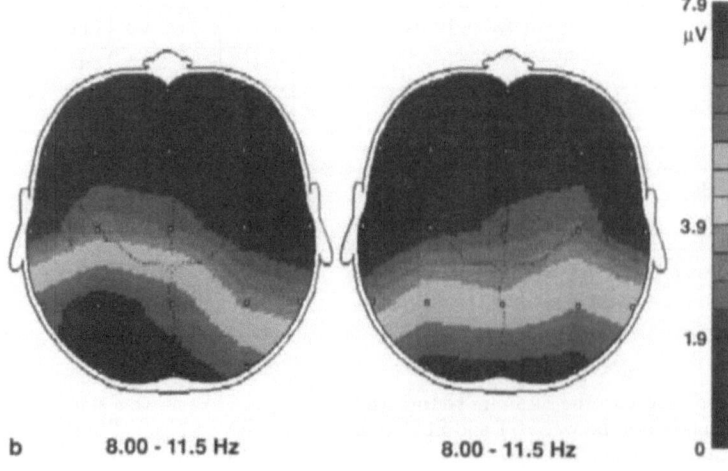

b   8.00 - 11.5 Hz   8.00 - 11.5 Hz

Fig. 40 a, b. Maps of flash-VEPs for P1 in the patient of Fig. 39. a Before removal of the tumor an asymmetry was seen with an amplitude reduction of more than 50% at O2. b Two months after removal of the meningeoma map features normalized totally

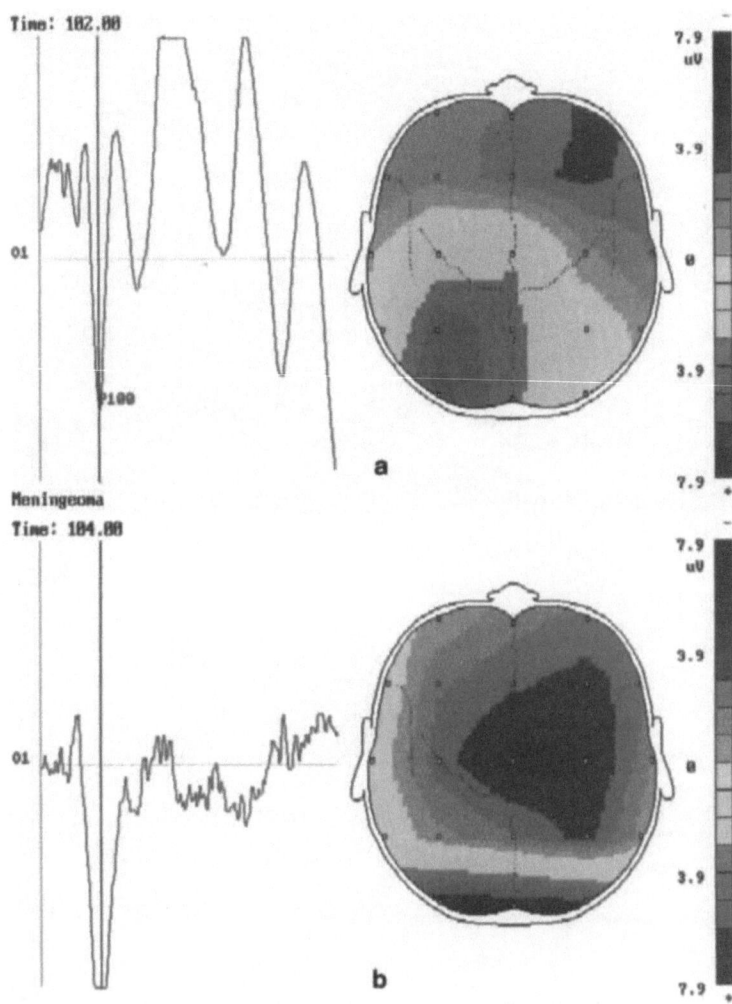

Hamburger (1989) described an extraordinary case of mapping results in a patient with a meningeoma (Figs. 39, 40). CT showed a large right occipital meningeoma, which was removed neurosurgically. Altered features of EEG and visual EP became fully normal during follow-up examinations. This case shows the advantage of two simultaneous neurophysiological methods in focusing a lesion and underscores the need to combine the EEG method with an EP test. The choice of EP method depends upon the affected brain area (e. g., VEPs for suspected occipital lesions, SEP for central and thalamic and AEPs for temporal lesions). EP maps open a new area in CNS diagnosis beyond the earlier recordings using only a few electrodes (Maurer et al. 1989 b).

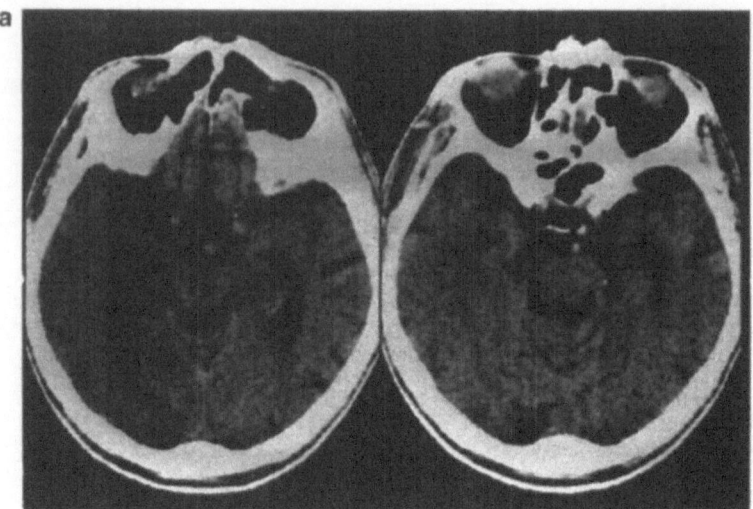

Fig. 41 a–c. CT and EEG
maps from a patient with a
left sided ischemic insult.
a CT with indication of a left
sided posterior insult. b EEG
maps in the four main frequency ranges, with an increase of delta and theta activity on the left side. c Corresponding z statistic map,
showing a greater deviation
from controls on left frontal
areas

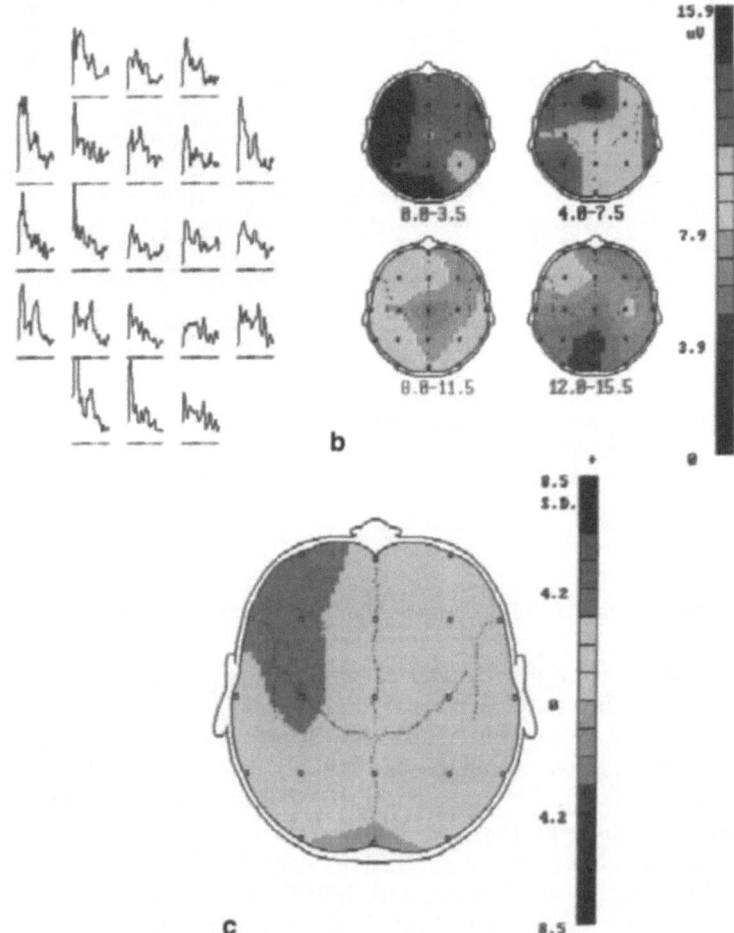

9.3.2.2 Cerebrovascular Diseases

Vascular processes and their corresponding EEG maps and findings have been investigated by Nagata (1984, 1989), Yamada et al. (1985), Nuwer et al. (1987), Ahne et al. (1988), Lechner et al. (1989), and by us in a series of patients suffering from transient ischemic attacks and strokes. Mapping methods are of special value in cases showing normal findings in structural methods such as CT and NMI (e. g. transient ischemic attacks). The noninvasiveness of mapping allows follow up studies to be carried out as often as are needed. Examples of map features and their CT equivalents are shown in Fig. 41.

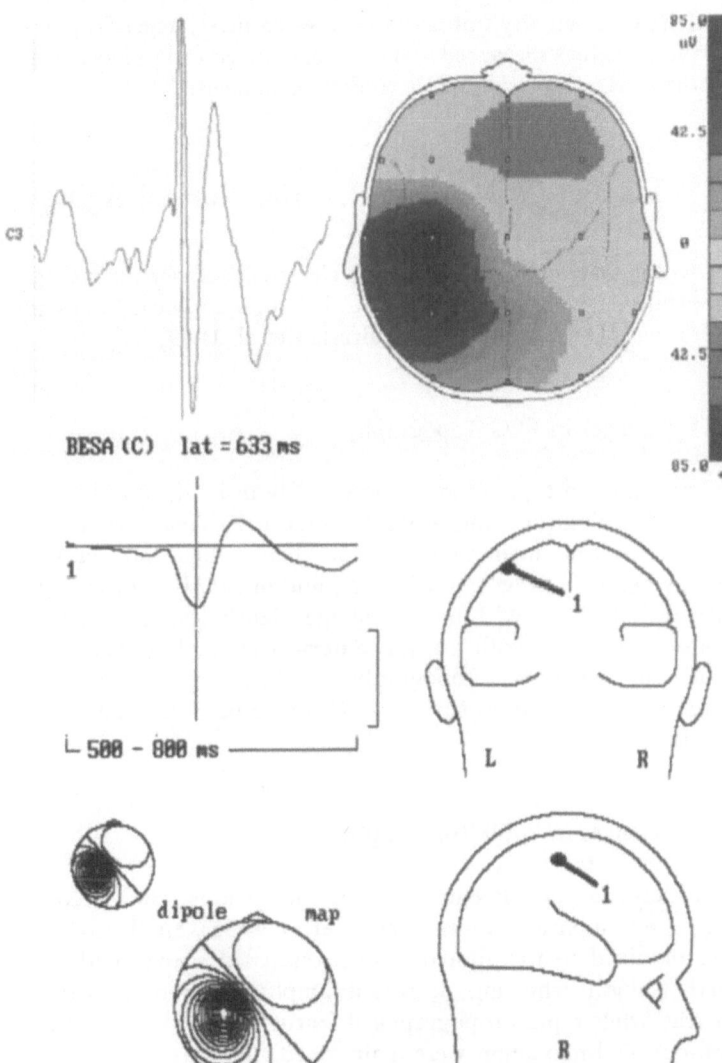

Fig. 42. Maps of transients recorded in a 6 year old boy with diagnosis of elective mutism. Detection of a spike (time domain) with a maximal negativity at cerebral points C3 and P3. Corresponding equivalent dipole calculation revealed a source of spike generation near electrode site C3 with a direction towards deeper brain structures

9.3.3 EEG Mapping of Transients

In epilepsy, conventional EEG may be superior to EEG mapping since the transformation process from the time to the frequency and spatial domains leads to a loss of specific EEG patterns. However, if epileptic discharges occur, mapping of raw EEG amplitudes and corresponding dipole calculations permit the identification of intracerebral sites where an increased excitability primary occurs (Lemieux et al. 1984; Lemieux and Blume 1986; Thickbroom et al. 1986; Harner et al. 1987; Nuwer 1988 a; Wong and Gregory 1988; Berger et al. 1989; Rodin and Cornellier 1989). Figure 42 shows spike discharges in a patient with an epilepsy. The focal aspect of the transients was underlined by taking into account the calculation of the underlying equivalent dipole. The epileptic source is not always at the site of maximum positivity or negativity but is often between these, depending upon voltage values measured at neighboring electrode placements and the direction of the dipole (radial or tangential).

9.4 EEG and EP Mapping During Normal Aging

Due to the systematic alterations of electrical activity during aging, group data are presented to display topographical EEG and EP features (Duffy et al. 1984 a; Breslau et al. 1989).

9.4.1 Changes in EEG Topography

EEG activity and topography undergo only insignificant changes during physiological aging processes. Figure 43 shows frequency maps of the four main ranges (delta, theta, alpha and beta) in young subjects (average age 27 years) and in elderly subjects (average age 74 years). With increasing age slightly less alpha activity is seen to occur, with a slight anteriorization; however these changes in activity and topography in old age were shown by confirmatory procedures such as ANOVA to be statistically nonsignificant.

9.4.2 Changes in P300 Topography

With increasing age, latencies of the endogenous P300 become longer and amplitude lower (Hegerl et al. 1985). Similar results were obtained in the control group discussed above, with the elderly subjects showing decreased amplitudes and prolonged latencies while typical topographical features with a parieto-temporo-occipital maximum were maintained.

Fig. 43 a, b. Topographical display of EEG frequencies in a sample of young healthy volunteers (a; n = 20; mean age, 27 years) and sample of geriatric controls (b; n = 14; mean age, 74 years). Using confirmatory statistical procedures, no significant differences were observed

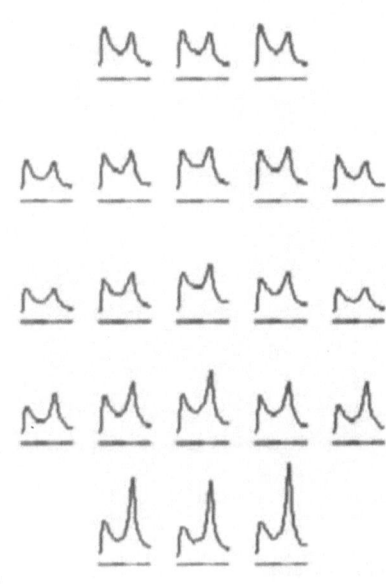
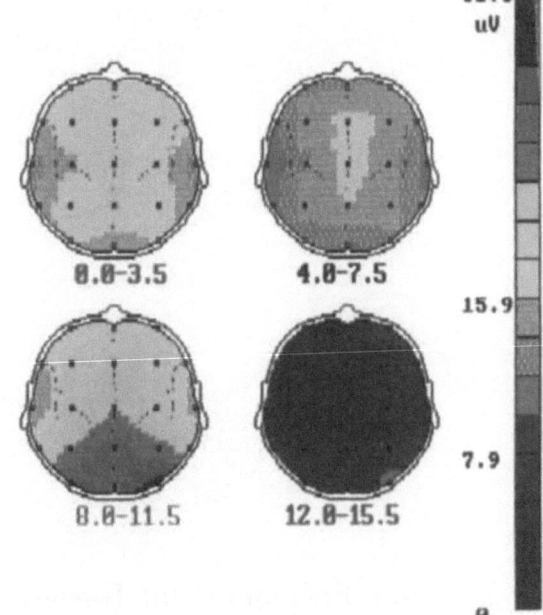

a

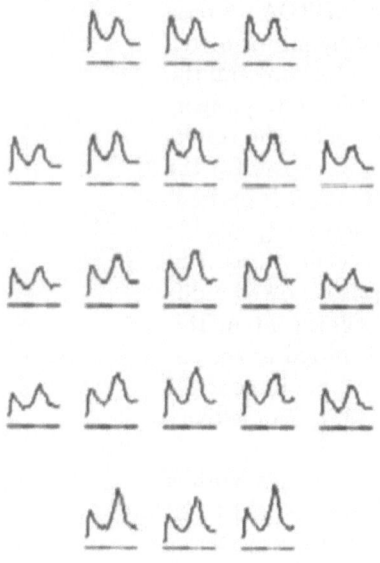
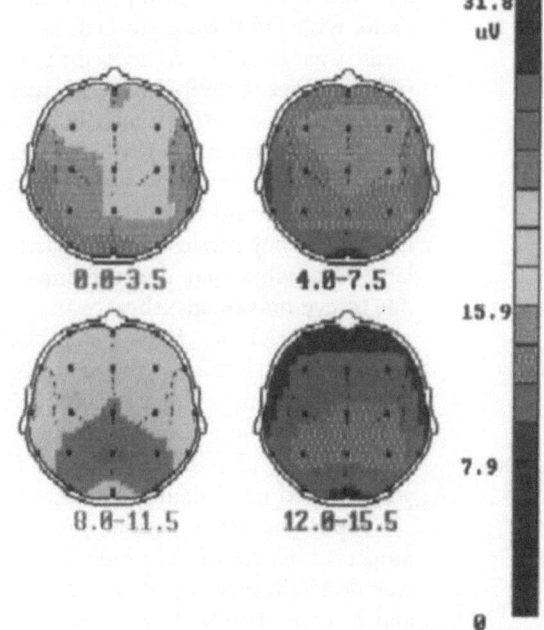

b

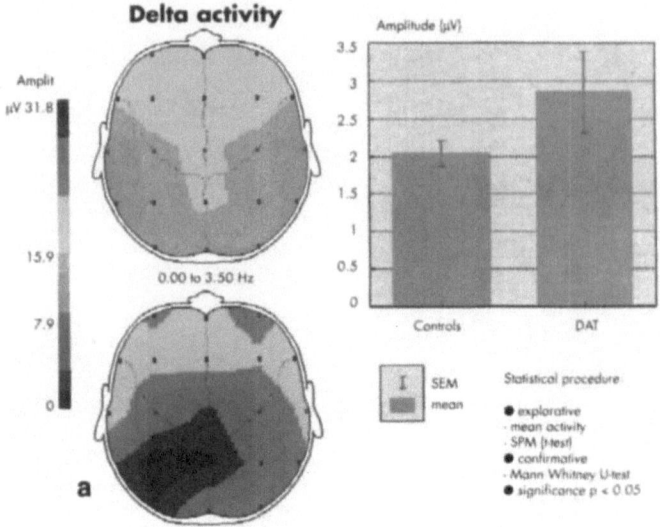

9.5 EEG and P300 Topography in Dementia of Alzheimer Type

Topographical data of EEG and EPs in cases of dementia of Alzheimer type (DAT) have been described by Duffy et al. (1984 b), Ihl et al. (1989), and Dierks et al. (1991). Due to the systematic alterations in DAT group data are presented first. Twenty patients with DAT were studied, (average age 75 years). The diagnosis was established according to NINCDS/ADRDA criteria (McKhann et al. 1984). Severity was determined by means of the Brief Cognitive Rating Scale (BCRS; Reisberg et al. 1983) and the cognitive performance test (SKT; Erzigkeit 1977). Only patients with moderate dementia were included in this study. The results from controls and from DAT patients are given in Fig. 44 for the four frequency ranges. With regard to delta activity, it is particularly noticeable that apart from the frontal increase (probably due to eye movement) there was an increase especially in parietal activity in DAT ($p < 0.05$ at Pz). Theta activity also showed a diffuse increase, however the increase was not significant in this group of patients. Alpha activity decreased and shifted to the anterior. The reduction in main frequency from 9.8 Hz in controls to 8 Hz in DAT patients was significant ($p < 0.05$). Beta activity was reduced in demented patients ($p < 0.05$).

From the same patient group as mentioned above with a moderate degree of dementia a testing of auditory elicited P300 was done (Maurer et al. 1988 a, c; Dierks et al. 1989 a; Dierks and Maurer 1990). As shown in Fig. 45, the demented patients exhibited higher amplitudes frontally, no difference centrally and an amplitude decrement in temporoparietal areas. Only the in-

Fig. 44 a–d. Topographic EEG maps. *Upper left,* geriatric controls; *lower left,* DAT patients for the respective range; *upper right,* bar diagrams, *lower right,* statistical procedure. a The increase in delta activity in the frontal and parietal regions of DAT patients was significant at the 5% level. b In the theta range one saw a tendency toward increase in DAT patients compared to geriatric controls. c In normal persons, occipitally structured alpha field; this was shifted toward the central field in DAT patients. Decrease in peak frequency from 9.8 Hz to 8.0 Hz. d In the beta range, decrease and anterior shift in DAT patients (lower head format)

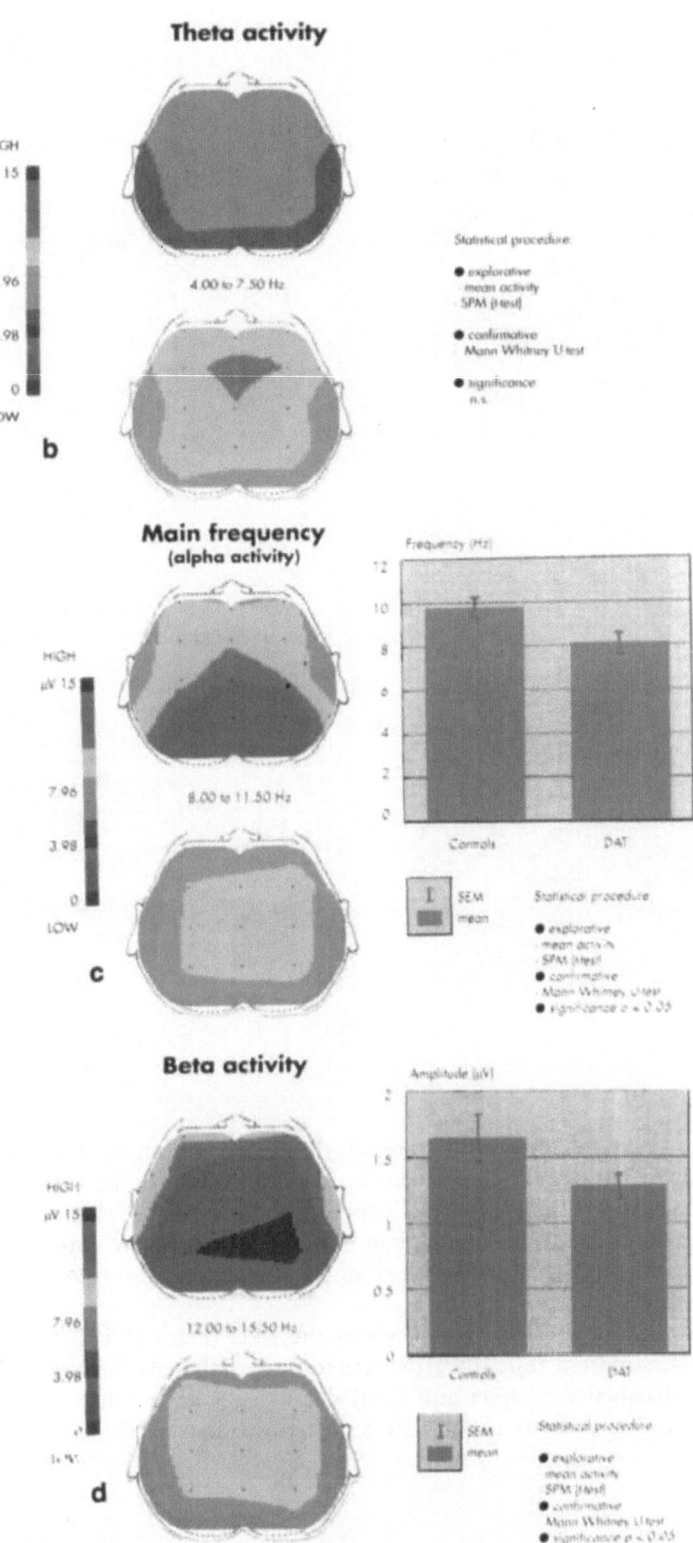

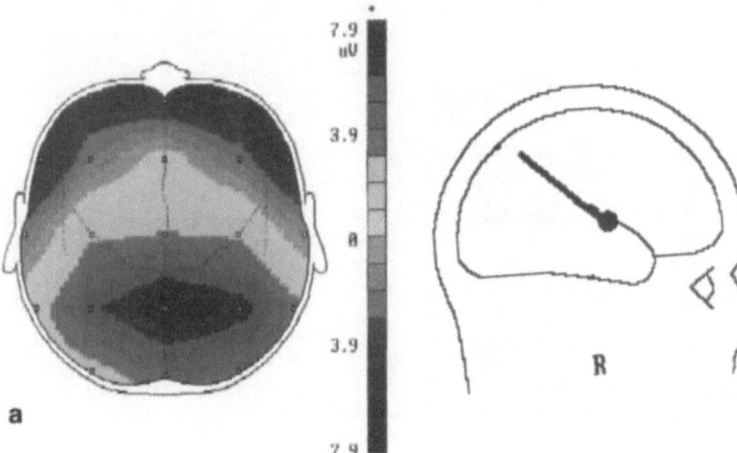

Fig. 45 a, b. Topographical display of P300. **a** Control group. **b** In DAT patients, a parietal maximum in P300 positivity and a displacement of P300 positivity toward frontal structures (the calibration bar has been altered to emphasize topographical features). *Right,* the equivalent dipole estimated according to the spatiotemporal dipole model of Scherg and von Cramon (1985)

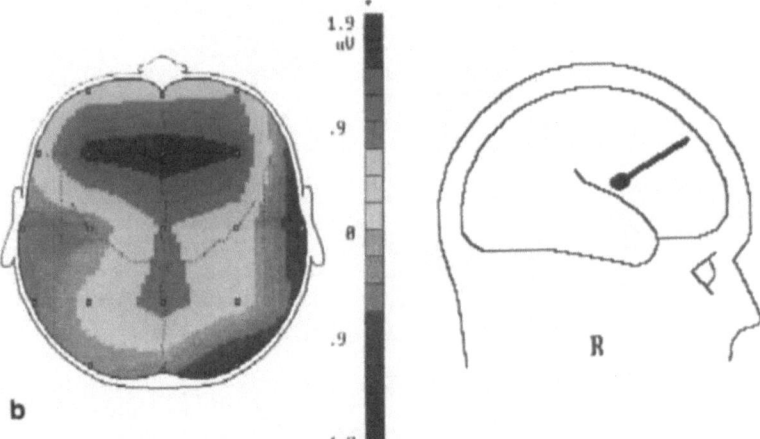

crease in bifrontal P300 amplitude achieved statistical significance. The frontal displacement of the P300 can best be explained by assuming an allocortical dipole which changes its direction due to cell damage in structures such as the entorhinal region, amygdala, and hippocampal formation (Maurer et al. 1989 d).

Generally slow frequencies in the EEG (delta and theta) increased with a shift towards parieto-temporal areas whereas faster frequencies (alpha and beta) decreased with a centrofrontal shift. P300 results underwent a shift towards frontal areas.

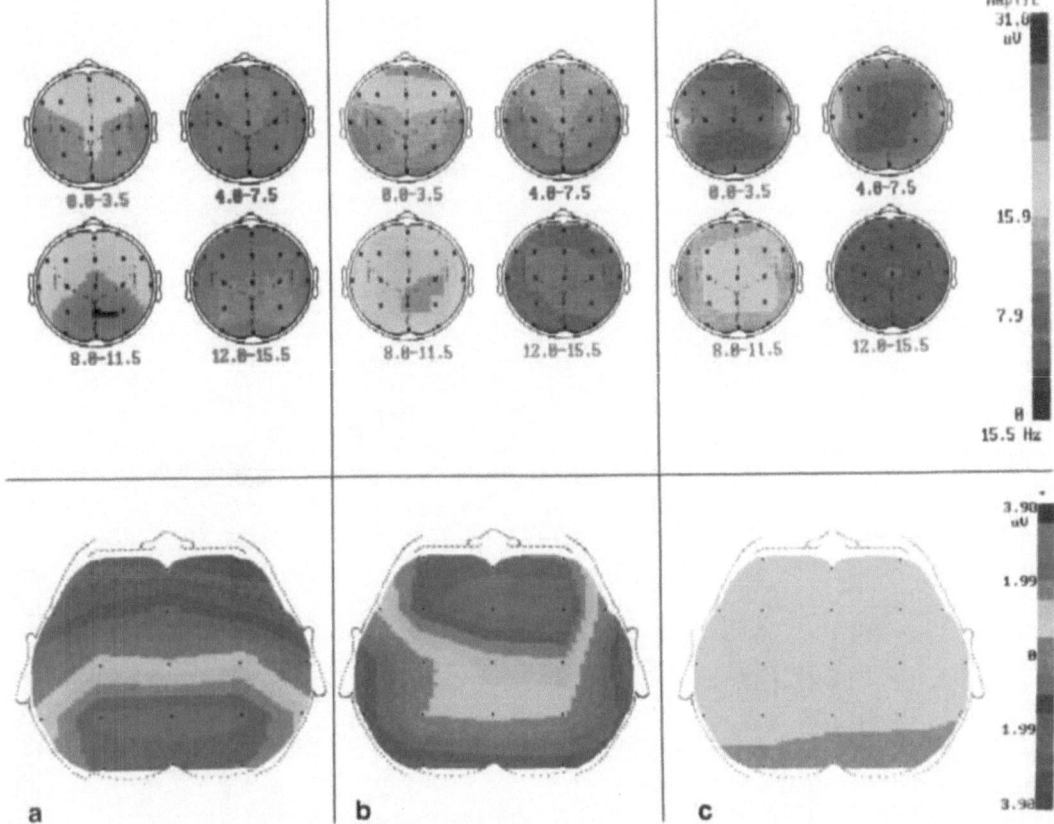

Fig. 46 a–c. Topographical display of stage dependent alteration of EEG activity and P300 fields. a Normal properties. b EEG and P300 topography of moderately demented patients. c Topographical features of severly demented patients. P300 has been recorded with a common average reference

9.5.1 Stage-Dependent Alterations of EEG and P300 Mapping in Dementia of Alzheimer Type

A series of 20 DAT patients was studied to investigate stage dependent alterations (Ihl et al. 1989). The patients were classified on the basis of BCRS into moderate or severely demented groups. Results of EEG and P300 mapping in these groups are shown in Fig. 46.

9.5.2 Differential Diagnosis of Dementia

Rare conditions such as dementive syndromes due to a luetic infection (progressive paralysis), Pick's disease and disturbed copper metabolism (Wilson's disease) are presented in terms of single cases. Group studies were conducted for the differential diagnosis of disease entities such as Parkinson's disease (PD), multi-infarct dementia (MID), DAT, PD with dementia (PDD), and major depressive disorders (MDD).

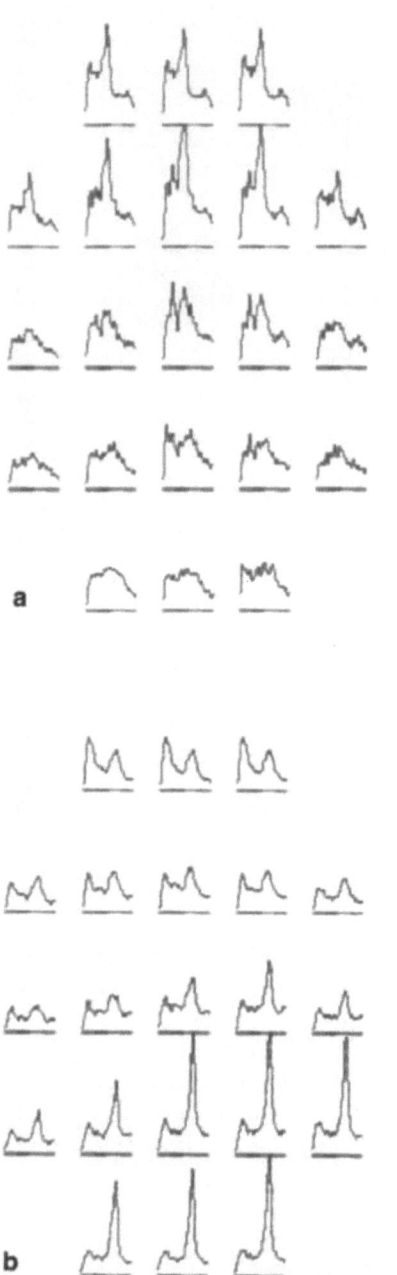

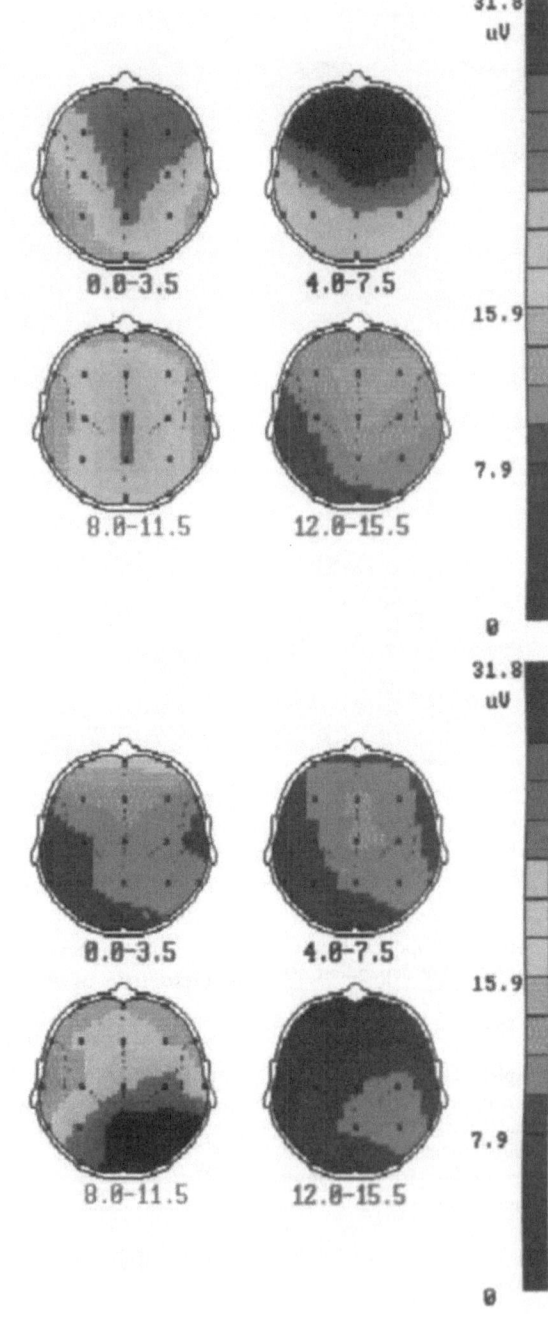

◀ Fig. 47 a, b. Topographic display of EEG activity in a patient with neurolues. **a** In the untreated state a considerable increase in theta and a decrease in alpha activity in combination with an anteriorization. **b** After treatment with penicillin alpha activity normalized and the increased theta disappeared

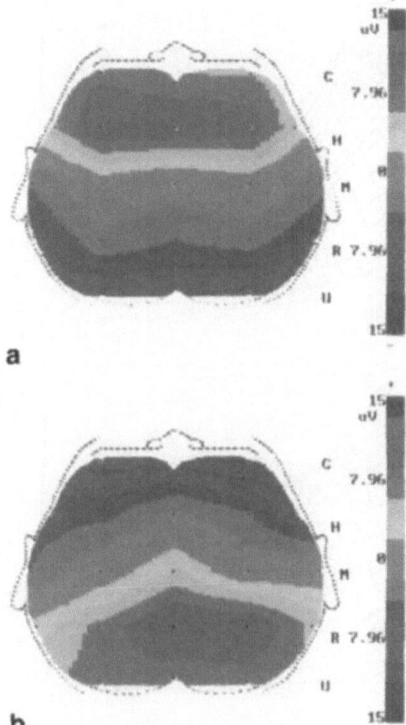

Fig. 48 a, b. Topographic display of P300 amplitudes in a case with DAT (a) and in a case with Pick's disease (b) ▶

9.5.2.1 Luetic Infection (Progressive Paralysis)

A patient with neurolues showed a typical EEG pattern with a frontal increase in slow activity (theta) before therapy. After therapy with penicillin a total normalization of the formerly described alterations took place (Fig. 47).

9.5.2.2 Pick's Disease

Compared to those in an age matched control group, a patient suffering from Pick's disease exhibited a significant increase in slow EEG activity in anterior parts of the brain, while P300 recording revealed a normal waveform and topography (Fig. 48).

9.5.2.3 Wilson's Disease

A patient with symptoms typical for Wilson's disease showed P300 topographical alterations characteristic of a dementive process (Maurer et al. 1988). Treatment with d-penicillamine led to the normalization of the formerly altered P300 (Fig. 49).

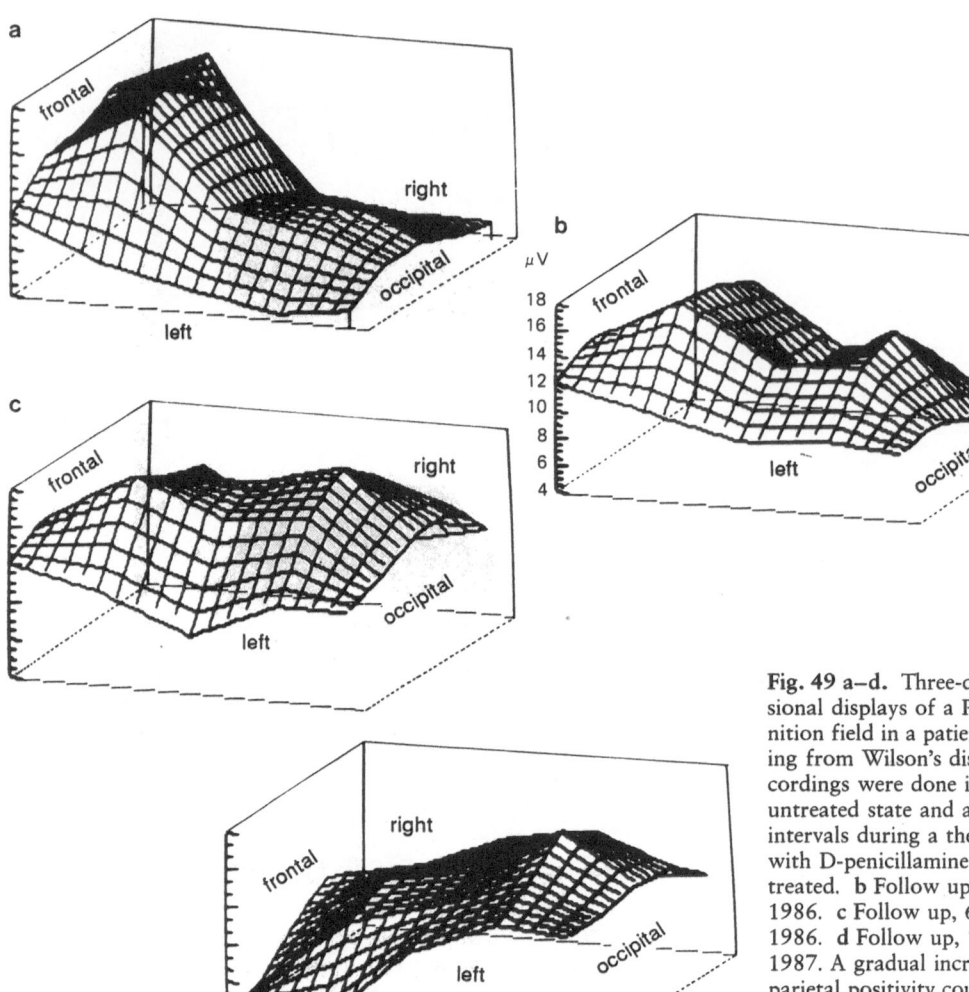

Fig. 49 a–d. Three-dimensional displays of a P300 cognition field in a patient suffering from Wilson's disease. Recordings were done in the untreated state and at regular intervals during a therapy with D-penicillamine. **a** Untreated. **b** Follow up, 5 May 1986. **c** Follow up, 6 Aug 1986. **d** Follow up, 18 May 1987. A gradual increase in parietal positivity could be observed with bifrontal attenuation

9.5.2.4 Parkinson's Disease, Parkinson's Disease with Dementia, Dementia of Alzheimer Type, and Major Depressive Disorder

Figure 50 presents group results from an explorative study and corresponding t values of EEG and P300 mapping in cases of PD, PDD, DAT, and MDD. Six persons in each group were investigated for the purpose of differential diagnosis (Maurer et al. 1989 b). As expected, EEG-features in DAT and PDD were very similar; in PD there was normal electrical activity and in MDD supression of the alpha rhythm. The P300 topography was very reliable differentiating the disease entities exhibiting a frontal P300 in DAT, a normal one in PD and an amplitude decreased one in MDD.

Fig. 50. Representation of t ▶ values for EEG activity and P300 results in dementia of Alzheimer type (DAT), Parkinson's disease (PD) Parkinson's disease plus dementive syndrome (PDD), and major depressive disorder (MDD) compared to controls. Numbers within head formats represent t values. Negative t values indicate a decrease and positive an increase in activity in the diseased group. In the case of P300 the corresponding topographical fields are shown (+, positive P300 field; − decrement of P300)

	DAT	PD	PD+DS	MDD
Delta	4.1 4.3 3.2 4.4 3.7 4.5	n.s.	1.5 2.3 1.7 2.2	n.s.
Theta	4.0 4.2 4.4 4.5 4.1 4.6 4.1 4.4	-2.1 -2.0 -2.2	2.1 2.0 2.1 2.2 2.0 2.2 2.0 2.1 2.1 2.2 2.2 2.1 2.3 2.5 2.9 3.1	n.s.
Alpha	n.s.	n.s.	n.s.	-2.2 -2.6 -2.4 -3.7 2.7 -2.7 -4.0 -2.4 -2.7 -2.3
Beta	-2.8 -2.7 -2.0 -2.2 -2.8 -4.9 -2.7 -3.6	-2.3 -2.0 -2.6	n.s.	n.s.
P300	⊕	⊕	?	⊖

	DAT	MID	Significance
Delta	+ +	+	Trend
Theta	+ +	+ +	n.s.
Alpha	− −	− −	n.s.
Beta	− − −	− −	p<0.05
P300			p<0.05

Fig. 51. Comparison of EEG and P300 data in DAT and MID patients

Fig. 52 a, b. Summary maps ▶ of EEG activity between 0 and 20 Hz in a patient with a schizoaffective disorder. a (before treatment): frontal slowing due to eye movements and weak activity within the alpha range (10.0–11.0 Hz). b (during treatment with neuroleptics): frontal slowing was maintained; besides that normal alpha field in a frequency range (10.0–11.0 Hz)

9.5.2.5 Dementia of Alzheimer Type and Multi-infarct Dementia

For differential diagnosis Maurer et al. (1989 b) studied a collective of patients suffering from DAT and MID. Results are shown in Fig. 51. As expected, P300 topography was maintained in the MID cases and allowed a differentiation to DAT cases in which a frontal elevation occurred.

9.6 EEG and EP Mapping in Psychoses

Mapping results in psychiatry are presented first in single cases to demonstrate the correlations between disturbed psychic function and electrical activity. Results are then discussed in groups of patients with defined psychopathological pictures and either under controled medication or medication free.

9.6.1 Case Studies

9.6.1.1 Schizoaffective Disorder (DSM-III: 295.7)

A 20-year-old, right-handed woman had shown no history of previous mental illness. Family history and drug screening were negative. The patient experienced an acute onset of disease with catatonic (excitement/mutism) and paranoid symptoms (delusions of reference) and verbal hallucinations. Mood and behavior were disorganized (incoherence; inappropriate, silly affect). Results of CT and NRI were unremarkable. One series of EEG and P300 mapping was completed prior to the initiation of drug

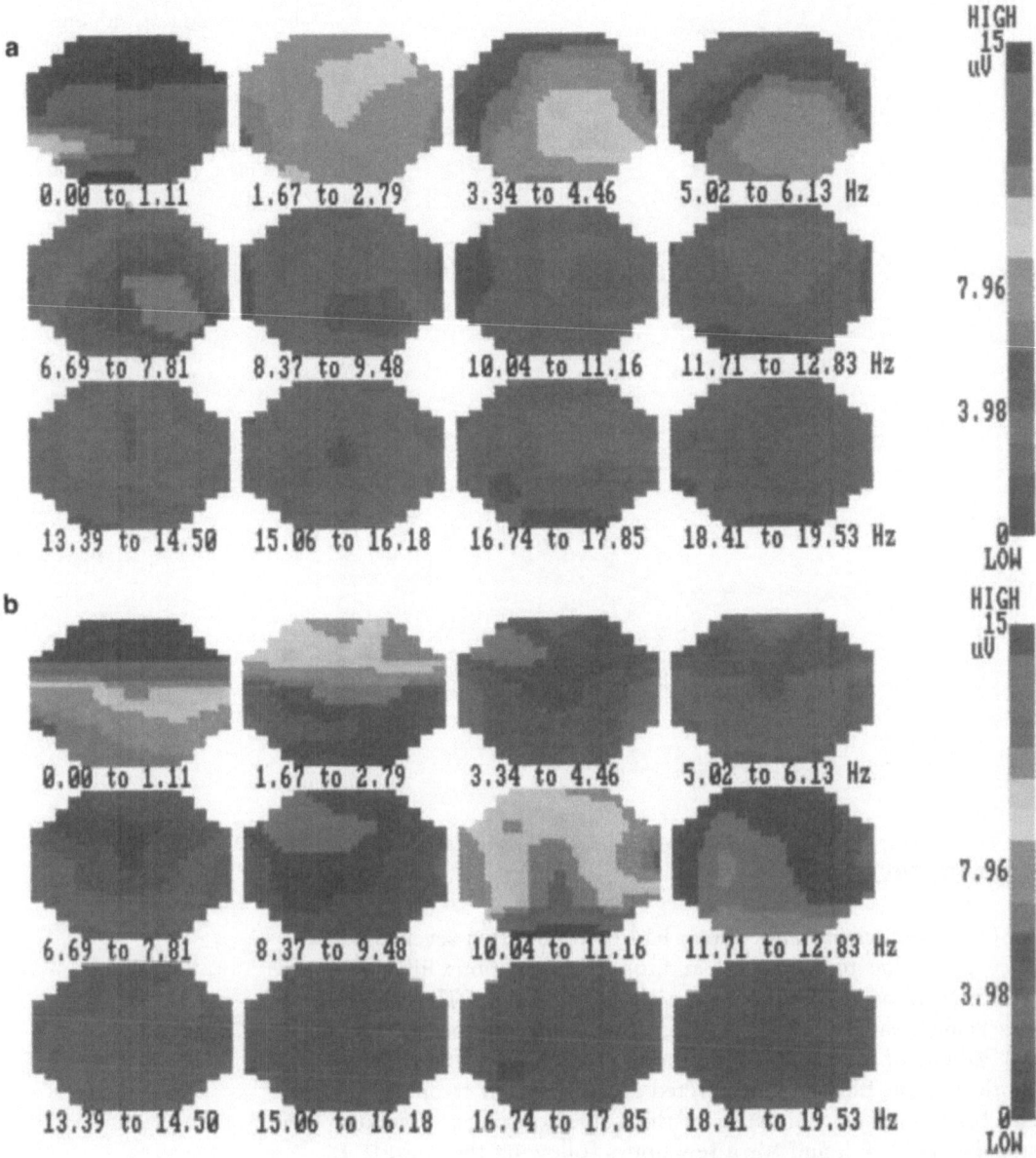

therapy and a second while the subject was receiving haloperidol and perazine. The premedication studies indicated an alpha reduction (Fig. 52) and a nearly abolished P300 response (Fig. 53). After therapy with neuropleptics activity within the alpha range normalized completely and P300 was normal, with a parieto-occipital maximum. The P300 counting task was performed accurately on both occasions.

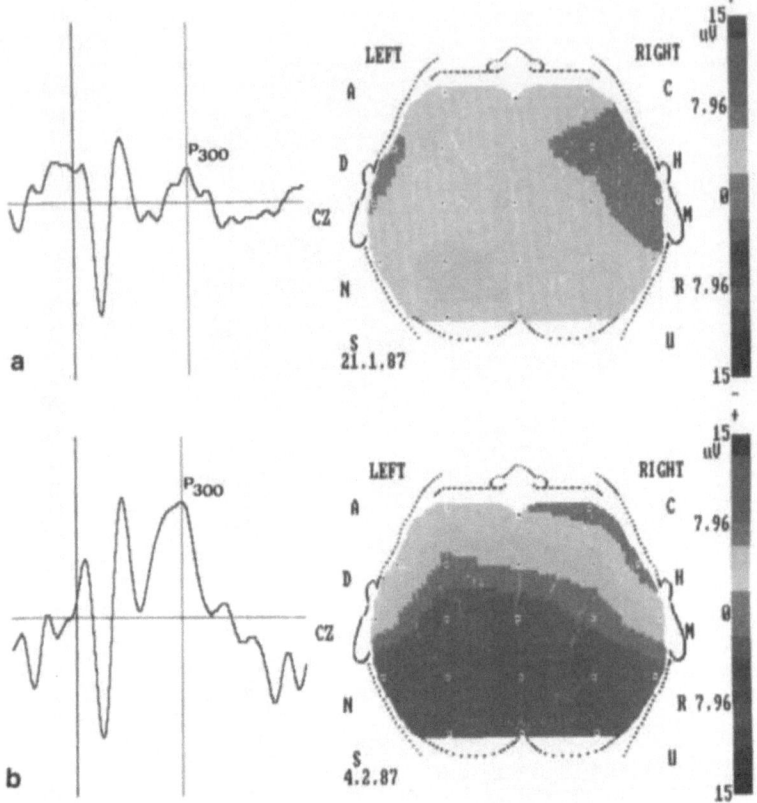

Fig. 53. a P300 topography before treatment with only a spurius P300 component at 380 ms. **b** P300 topography after treatment with a clearly definable P300 wave at cerebral points P3, Pz and P4 at 356 ms

9.6.1.2 Schizophrenic Disorder, Paranoid Subtype (DSM-III: 295.3)

A 30-year-old, right-handed man had been admitted seven times since 1979 due to delusions, auditory hallucinations, illusions, phonema, ideas of reference, and thought hearing. The subject was being treated with medication (Carbamazepin and Melperon). Results of CT and MRI were unremarkable. During the psychotic episode, EEG was interpreted as showing left temporal focus. Mapping was completed on this subject during a psychotic state lasting 2–3 h and for a few hours following the episode. It was also possible to compare the patient's data to baseline data obtained during a previous study. Baseline FFT data indicated an asymmetric alpha field with minimal beta range activity. FFT data from the EEG performed during the psychotic episode showed evidence of increased beta activity with left accentuation (Fig. 54 a). Approximately 2 h following the psychotic episode, the FFT was similar to the baseline (Fig. 54 b). The baseline P300 was poorly formed but replicable (Fig. 55 a). Also, the P300 latency response was abnormally prolonged. Even though the patient was able to perform the P300 task accurately, there was no

Fig. 54 a, b. Topographic display of EEG activity in a patient during a psychotic episode. **a** during the psychotic an increased beta activity at electrode sites T3 and C3 occurred. **b** Attenuation of the formerly described increased beta activity

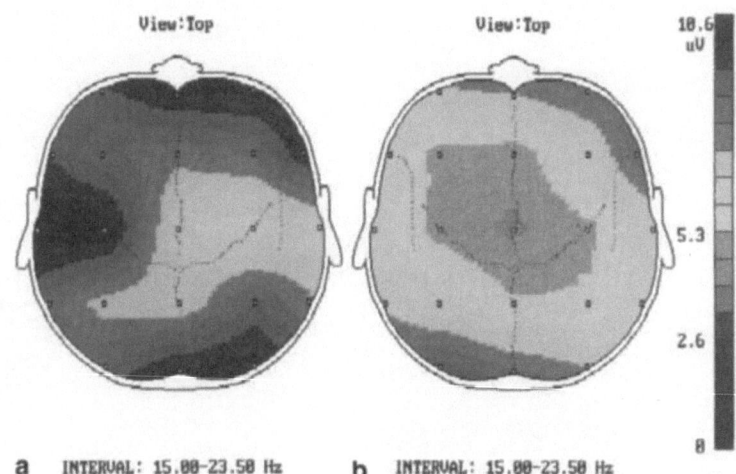

View:Top
View:Top
10.6
uV
5.3
2.6
0
a
INTERVAL: 15.00-23.50 Hz
b
INTERVAL: 15.00-23.50 Hz

Fig. 55 a–c. Topographic display of P300 activity in the patient mentioned in Fig. 54. **a** subject's baseline auditory P300 response. **b** subject's P300 response during the psychotic episode. **c** subject's P300 response 2 hours after the psychotic episode

PZ
a
7.9
uV
3.9
0
3.9
7.9
PZ
b
PZ
c

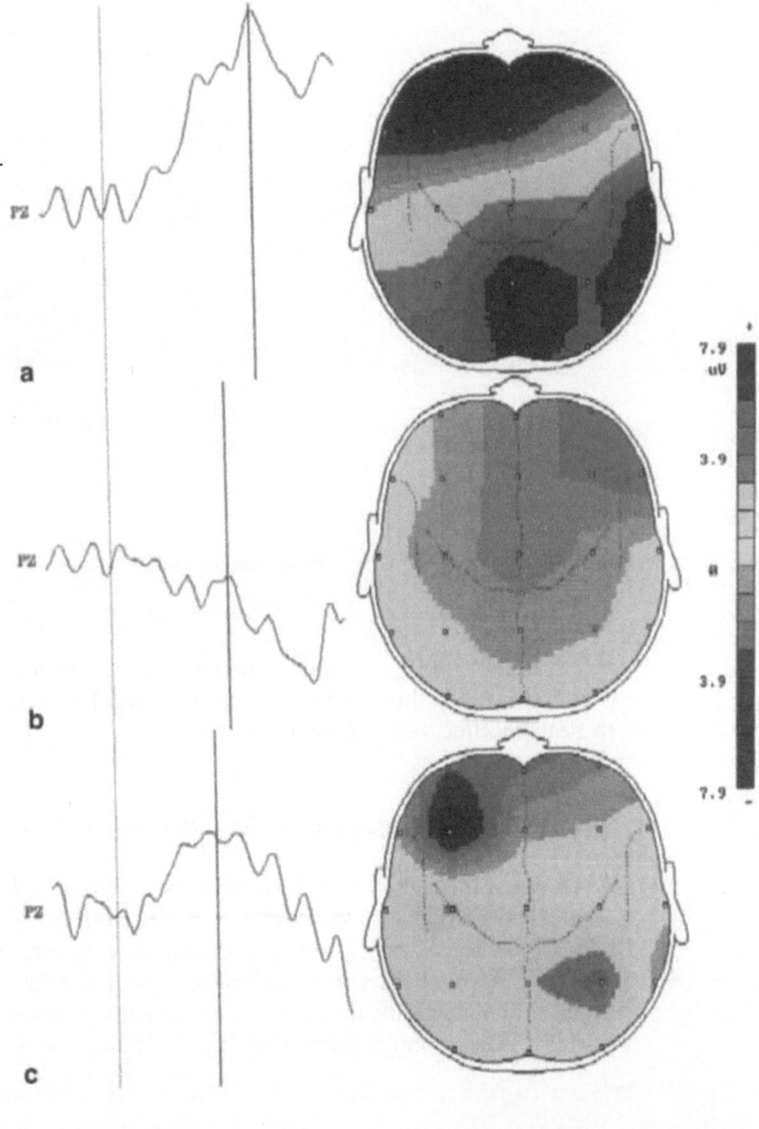

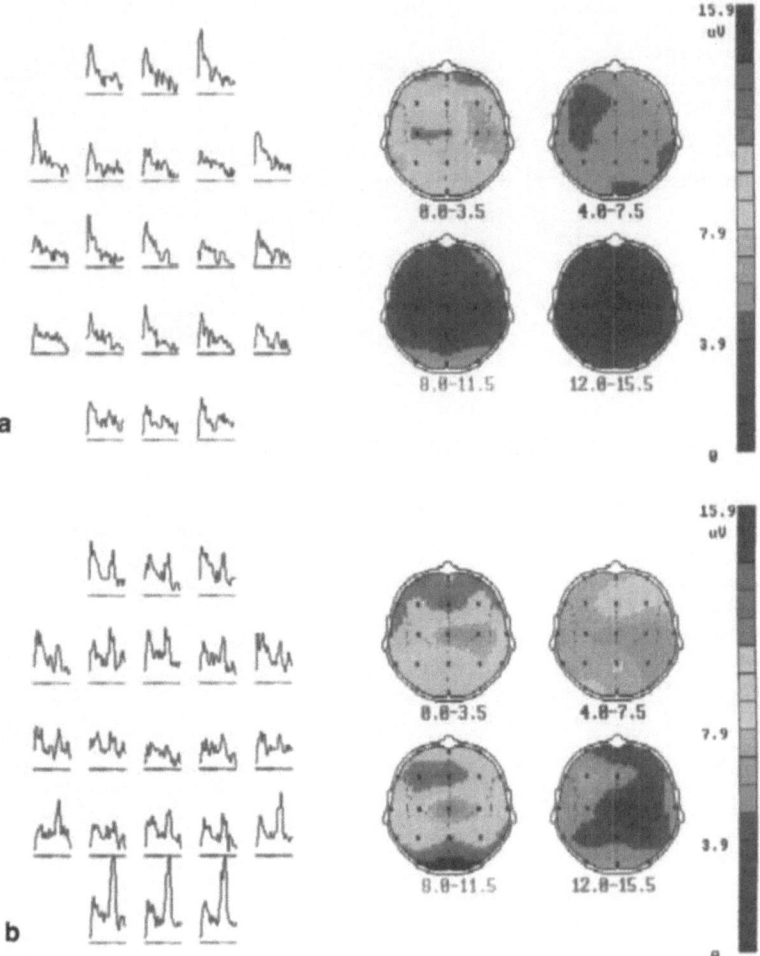

identifiable P300 response during the psychotic episode
(Fig. 55 b). A few hours after the episode, the P300 had returned
to near baseline values (Fig. 55 c).

9.6.1.3 Major Depressive Disorder (DSM-III; 296.2)

A 44-year-old, right handed woman experienced a long-lasting
depressive episode as a result of a personal conflict. Symptoms at
admission were depression, diminished interest, significant
weight loss, psychometric retardation, loss of energy, and diminished ability to think or concentrate. Suicide attempts were reported in the recent history. Results of CT and MRI were unre-

Fig. 56 a–d. Topographic display of EEG activity and P300 amplitudes in a patient suffering from major depression. **a** Reduced alpha activity prior to medication. **b** Increase in alpha activity and reorganization of the alpha field after continous antidepressant therapy. **c** Low amplitude P300 response prior to medication (common average reference). **d** Nearly normal P300 response after continuous therapy (common average reference)

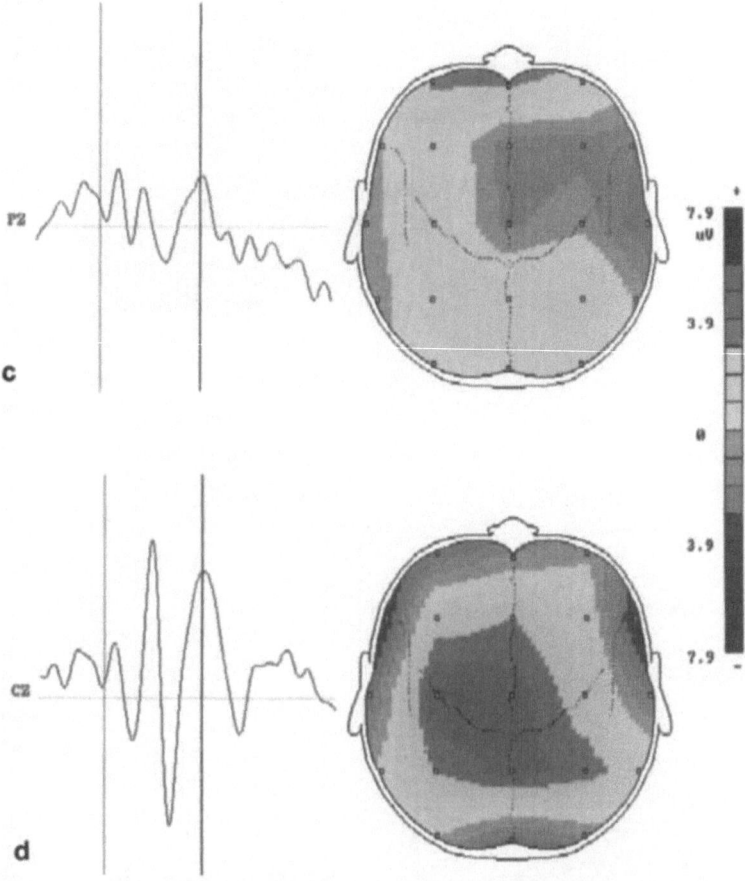

markable. Upon admission, the patient's Hamilton Score was 17 and her Adjective Mood Score was 31. Following antidepressant therapy, the Hamilton Score improved to 9 and the Adjective Mood Score decreased to 20. Treatment was with unspecified antidepressants. One mapping series was completed prior to the initiation of drug therapy and another while the subject was under medication. The premedication studies indicated reduced alpha activity and low amplitude P300 response (Fig. 56 a,c). After therapy with antidepressants there was a much improved alpha field in the occipital area and a nearly normal P300 response (Fig. 56 b, d). Every effort was made to ensure the subject's diligence during both test sessions; the P300 counting task was performed accurately on both occasions.

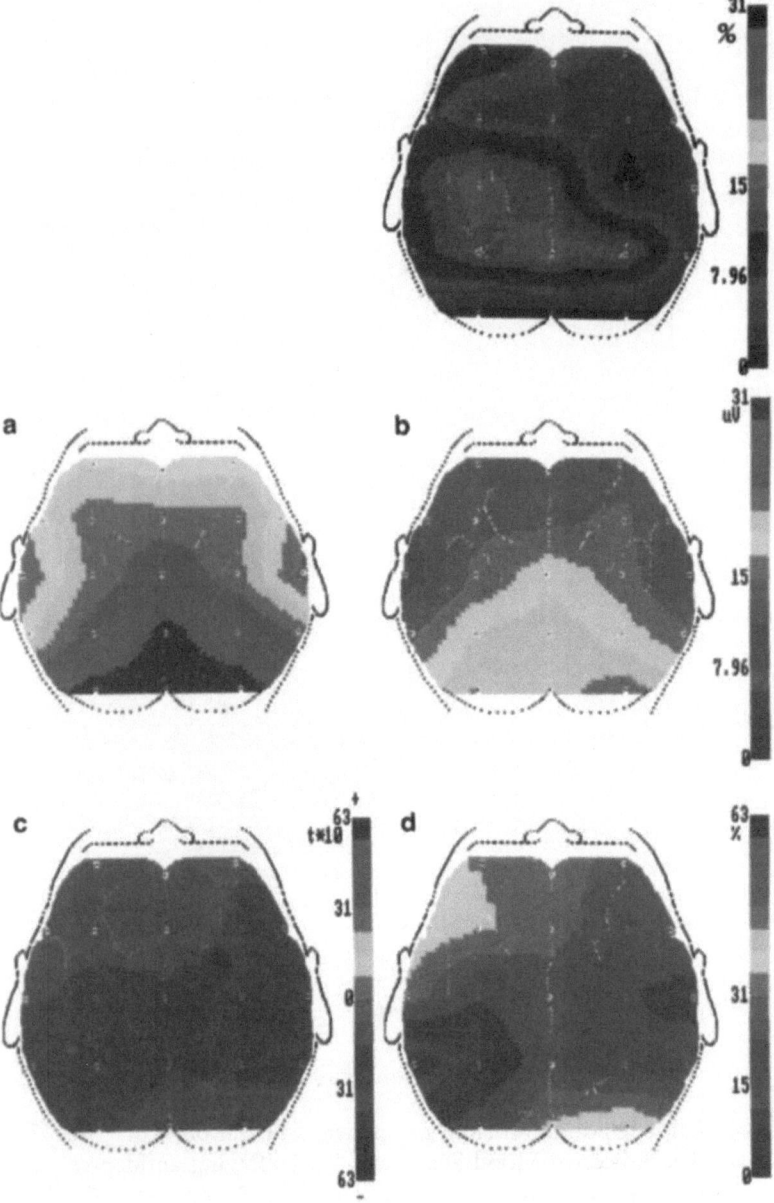

Fig. 57. Practical significance map for the delta band (0–3.5 Hz) gained by comparison of the control group with the hebephrenic subgroup. Maximal values are measured over the midline and left central and parietal electrodes, indicating an increase of delta activity in these regions

Fig. 58 a–d. Comparison between the controls and the hebephrenic subgroup for the alpha band (8.0–11.5 Hz) in the eyes closed state. **a** Mean value for 10 normals, **b** mean value for 10 hebephrenics, **c** a t value map exhibiting the difference between controls and hebephrenics. The color scale indicates t values multiplied by the factor 10. Positive t values mean augmented and negative t values decreased activity in hebephrenics compared to controls. **d** Practical significance map indicating the clinical relevance of the findings. The map was constructed using t values

9.6.2 Group Results

9.6.2.1 EEG Mapping in Schizophrenia

Early results of mapping in cases of schizophrenia have been described by Buchsbaum et al. (1982 a, 1986), Morihisa et al. (1983, 1985), Morstyn et al. (1983 b), Guenther and Breitling (1985), and Karson et al. (1987). The goal of a pilot study carried out by Dierks et al. (1989) and Maurer et al. (1989 b) was to divide a series of 30 schizophrenics by topographical mapping into groups according to DSM-III and ICD-9 criteria and to delineate their EEG patterns in the delta, theta, alpha, and beta ranges and P300 results. Several test procedures, including F test, Student's t test and the test of practical significance, were used to compare control data with patient data.

There were higher amplitude values in the delta range only in the hebephrenic subgroup in the eyes-closed state, with a predominance centrally and parietally and a slight accentuation on the left hemisphere (Fig. 57). The theta band showed a decrease of amplitude values occipitally in the paranoid group in the eyes-closed state but an increase in the eyes-open state in the hebephrenic and paranoid groups. Our findings of decreased alpha activity were most obvious and consistent in the eyes-closed state, especially in the hebephrenic group (Fig. 58) but this also characterized the paranoid and the residual groups (Fig. 59). In the eyes-open state a significant increase was seen with an accentuation in all three groups. For the beta bands the same alterations occurred as in the alpha range, i. e., a decrease in the eyes closed state and an increase in the eyes-open state. Remarkable was an

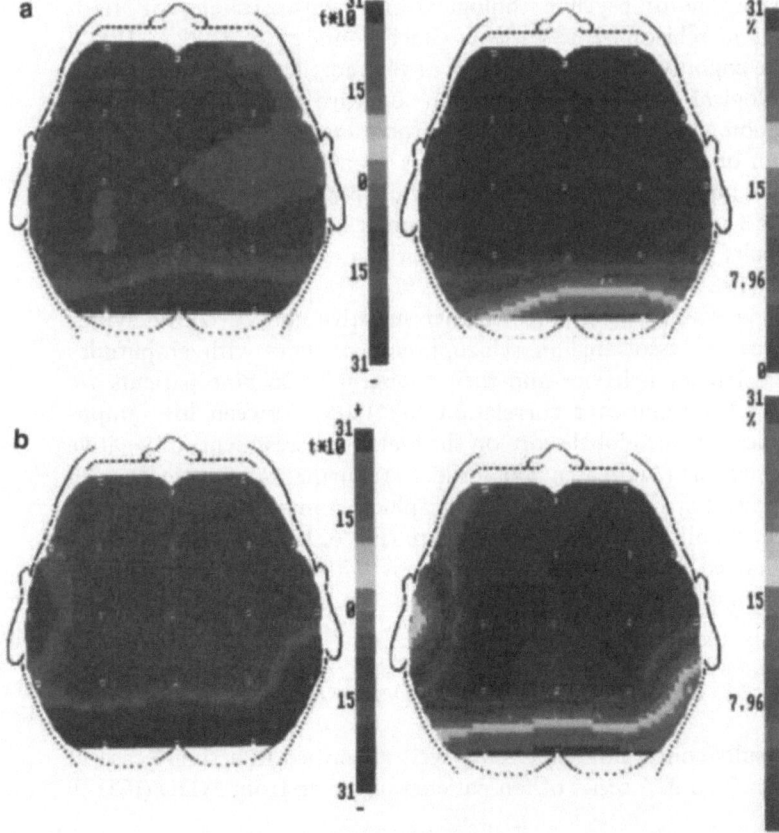

Fig. 59 a, b. *Left: t* value maps; *right:* practical significance maps. a Results of comparison between the controls and the paranoid subgroup for the alpha band (8.0–11.5 Hz) in the eyes-closed state, indicating an occipital decrease. b Results of comparison between the controls and the residual sugroup for the alpha band (8.0–11.5 Hz) in the eyes-closed state indicating an occipital decrease

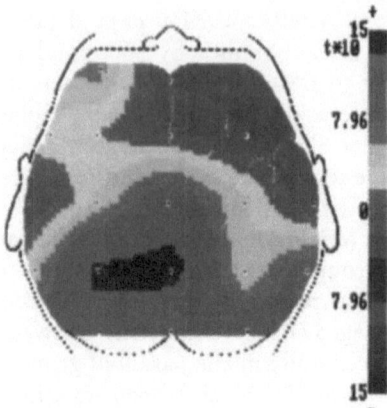

Fig. 60. A t value map indicating the difference between the controls and the hebephrenic subgroup for the beta$_1$ band (12.0–15.5 Hz) in the eyes-open state

increase of activity in the slow beta band (12–15.5 Hz) in the eyes-open state with a maximum over Pz and P3 in the hebephrenic group (Fig. 60).

9.6.2.2 P300 Mapping in Schizophrenia

Disurbances of information processing play a central role in explanations of psychopathological and neuropsychological findings in schizophrenic patients (Buchsbaum et al. 1982 a, 1986). The cognitive wave P300 is one of the most important neurophysiological variables, depending on cognitive processes of discrimination and thus on information processing. The amplitude reduction of P300 in schizophrenics has been found to be a sensitive but nonspecific indicator of schizophrenia (Morstyn et al. 1983 a, b; Faux et al. 1988; Shenton et al. 1989). Early mapping results have been published by Morstyn et al. (1983), Shenton et al. (1989) and Maurer et al. (1990). The purpose of our own study was to determine whether negative and psychotic symptoms are associated in schizophrenic patients with amplitudes and latency behavior and their topography. In nine patients we found a significant correlation ($p < 0.05$) between low amplitudes and the global score on the scale for assessment of negative symptoms (Andreasen 1982; Fig. 61). Furthermore patients with negative symptoms had a topographical minimum of the equivalent dipole in the right hemisphere (Fig. 62). Similar results were obtained under multilead conditions without mapping by Pfefferbaum et al. (1989).

9.6.2.3 EEG and EP Mapping in Depression

Results concerning depression were described in a single case in Sect. 9.6.1.3. A series of ten patients suffering from MDD (ICD-9:

Fig. 62 a–c. Topographical display of P300 curves gained in a normal person (a) a patient with a mild negative symptomatology (b) and a patient with pronounced negative symptoms (c)

Fig. 61. Correlation between P300 and the global score of the scale for assesement of negative symptoms (r = − 0.72; p < 0.05)

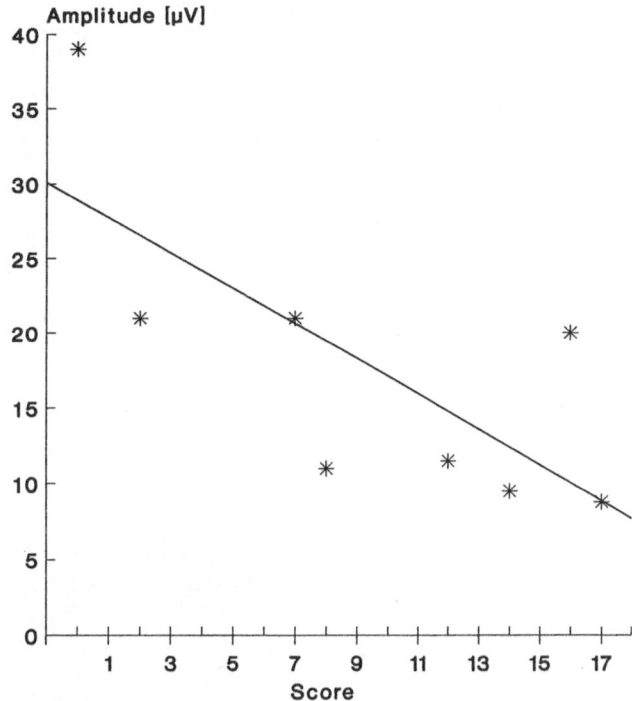

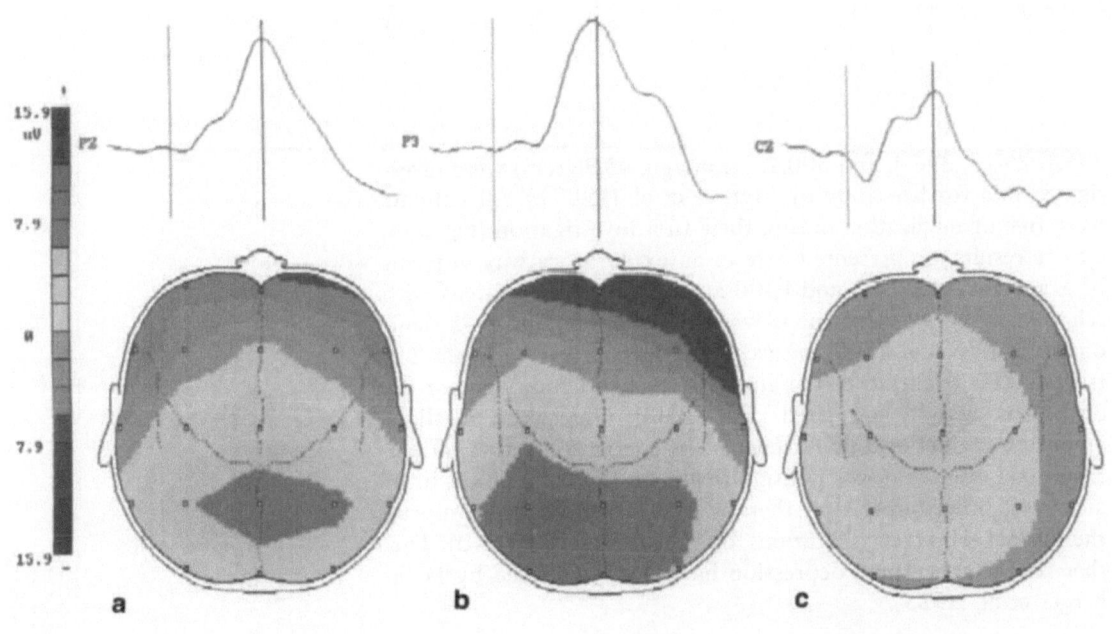

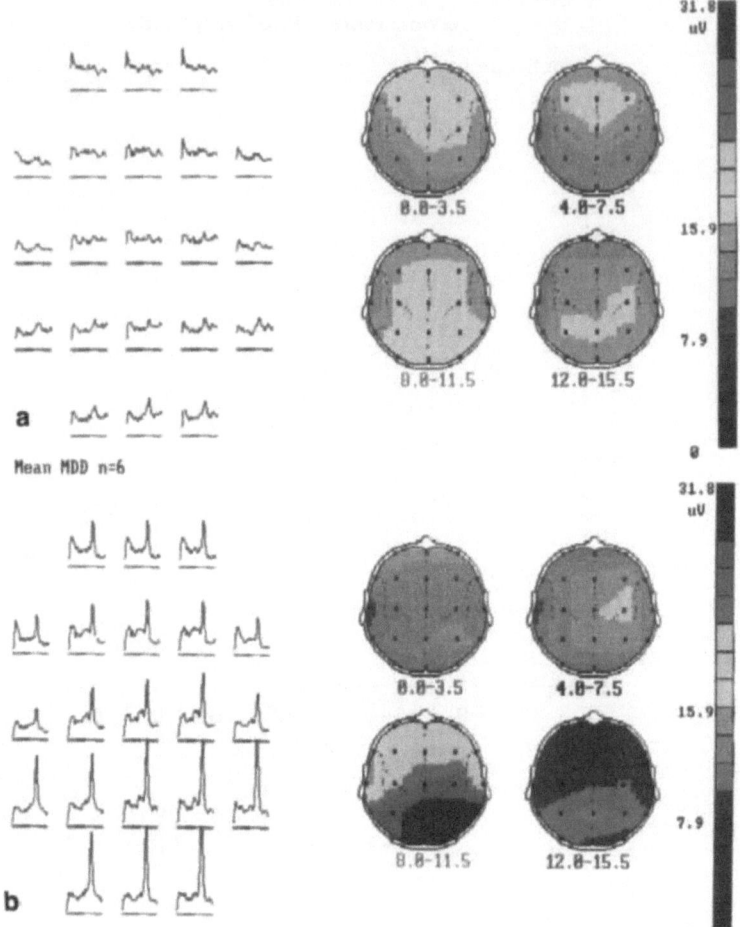

Fig. 63 a–d. Topographic P300 display in a group of patients suffering from major depression. a EEG topography in the four main frequency domains before treatment. b EEG topography after treatment with antidepressant drugs. c P300 topography before treatment with a normal P300 pattern, but slightly decreased parietal P300 amplitudes (Pz = 10.4 µV). d P300 topography after treatment with an increase of P300 amplitude in the parietal area where now 16 µV could be measured

296.1, 296.3, 296.4, and 300.4; mean age, 45.9 years) were investigated in a further study by Maurer et al. (1989 b). All patients were free of medication during their first investigation (Fig. 63 a, c). The results are presented here in an exploratory way in terms of grand mean of EEG and P300 activity. To explore areas with relevant changes t value maps where calculated. Figure 63 demonstrates EEG and P300 features. The only alteration in the untreated state occurred in the alpha frequency band, where a decrease of activity occurred. The P300 underwent a slight amplitude reduction but no change in topography; the parietal amplitude reduction was nonsignificant on the basis of confirmatory test procedures. After therapy with an antidepression drug the former observed alterations disappeared (Fig. 63 b, d). Further results concerning depression have been described by Pockberger et al. (1985).

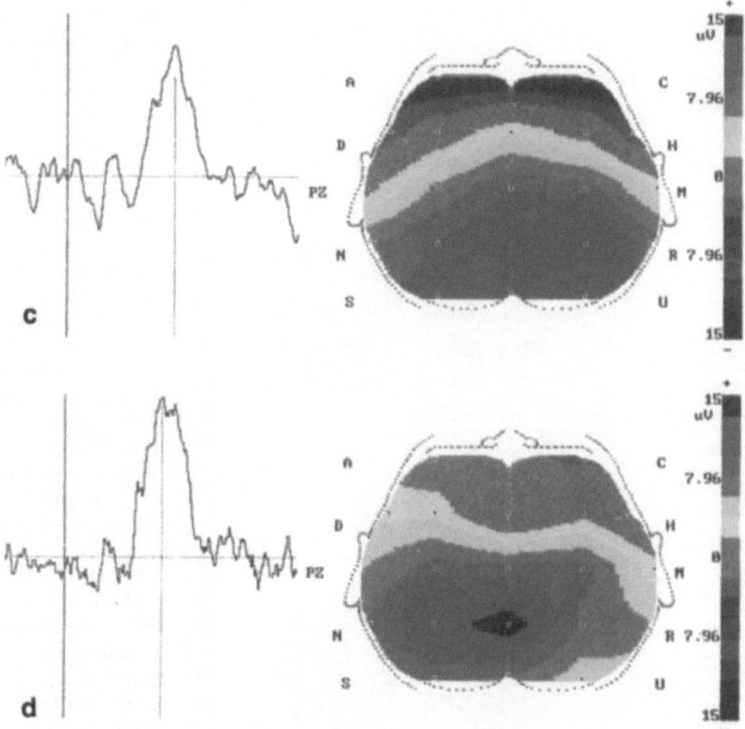

9.7 Mapping in Clinical Psychopharmacology EEG and EP

9.7.1 EEG Mapping After Application of Drugs

With drug therapy one must first consider the effects of the drugs themselves upon the neuronal activity of a healthy brain; these can be demonstrated in volunteers. Secondly, the effects must be considered which drugs exert upon an altered neuronal activity due to diseases (i. e. Alzheimer's disease).

The effects of several drugs upon a normal brain can also be convincingly investigated in patients undergoing anesthesia due to a disease not affecting the brain (Duffy et al. 1984 a; Kitani et al. 1985; Carl et al. 1989; Engelhardt et al. 1990). We studied various inhalational and intravenous agents off-line and present here an example of the short-acting barbiturate thiopentone (Engelhardt et al. 1989). Figure 64 a–c shows EEG maps of a 64-year-old woman during induction of anesthesia with thiopentone.

The effects of drugs upon an altered neuronal activity due to a disease can be demonstrated in Alzheimer's disease. The cognitive enhancing substance pyritinol diminished slow EEG activity and increased alpha activity accompagnied by corresponding to-

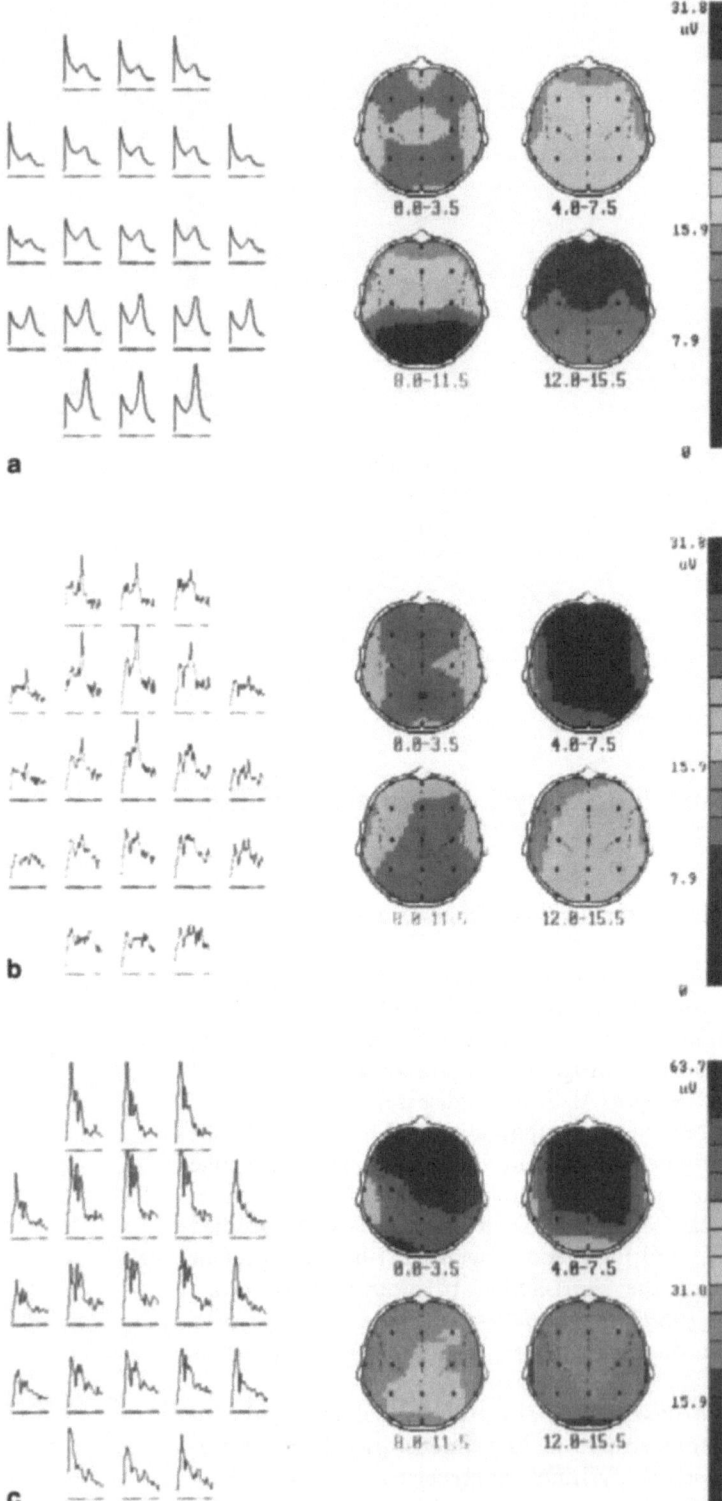

Fig. 64 a–c. Topographical display of EEG activity in the main frequency ranges before and during thiopentone injection. **a** Baseline recording before injection. Frontal delta activity was due to eye movements and blinks in the unpremedicated anxious individual. **b** 18 s after thiopentone injection. A slowing of the EEG with increased theta activity over central and parietal areas could be observed. Activities within the alpha and beta range also increased. **c** 28 s after thiopentone injection. EEG signs of anesthesia were maximal with high delta activity over frontal and theta activity over central areas. **d, e** Topographical display of EEG activity in the delta and alpha frequency ranges before and after administration of pyritinol. Delta activity in a geriatric control person (**e** 1), in a DAT patient (**e** 2), and in the same patient after drug ingestion (**e** 3). Alpha activity in a geriatric control person (**d** 1), in a DAT patient (**d** 2), and in the same patient after drug ingestion (**d** 3)

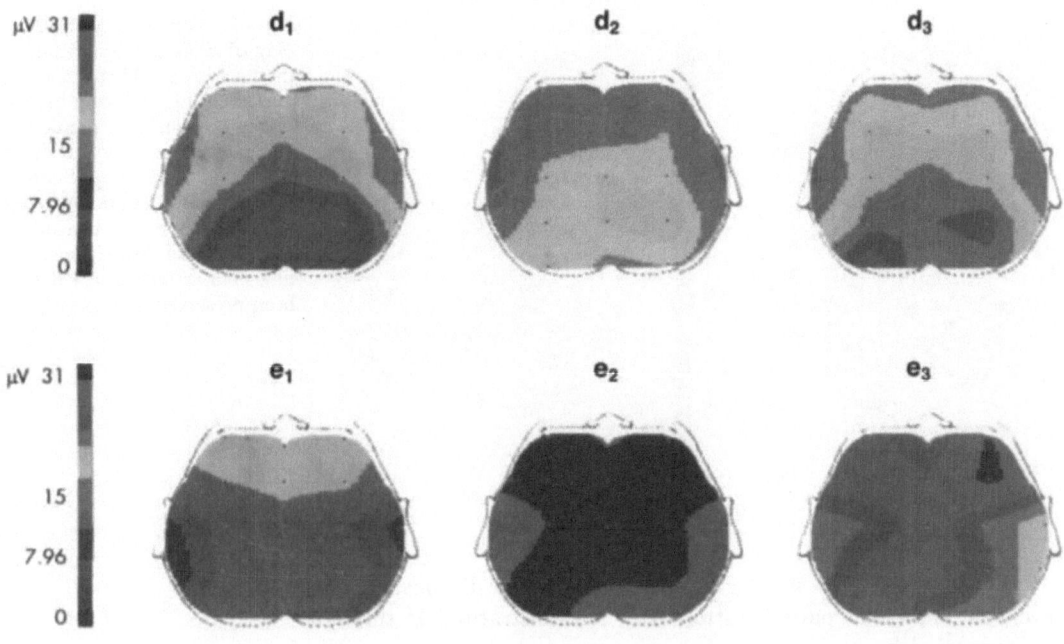

pographical shifts. Four hours after a single dose of 600 mg pyritinol dihydrochloride 1 H_2O the delta which could be diffusely recorded in all cerebral areas in the untreated state was reduced after drug administration (Fig. 64 d). The alpha activity, which without medication had been strongly reduced and shifted toward central, showed after treatment a reorganization in the occipital region with a simultaneous marked increase in activity (Fig. 64 e).

Effects of substances such as neuroleptics, thymoleptics, tranquillizers, lithium (Beaubernard et al. 1987; Thau et al. 1989), and cognitive enhancers have been extensively described in volunteers by means of mapping by Saletu et al. (1989) and others (Pockberger et al. 1984; Itil and Itil 1986; Anderer et al. 1987; Maurer et al. 1988 d; Fritze et al. 1989). These studies in volunteers show that quantitative EEG and EP methods and their topographical display can be used to determine cerebral bioavailability, dose-efficacy relations, bioequipotency and time-efficacy relations of substances. Figure 65 demonstrates that a tranquillizer provoked beta activity over frontocentral areas, and how this drug-induced increased activity was blocked by the benzodiazepine antagonist flumazenil. This example shows the superiority of mapping over earlier recordings with only a few channels in pharmacoelectroencephalographic studies.

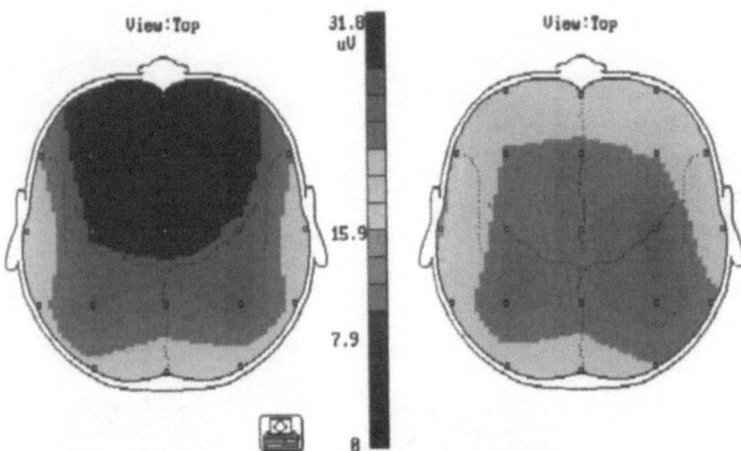

Fig. 65. Topographical display of EEG activity within the beta range (15.0–23.5 Hz) due to an injection of a benzodiazepine. Evidently increased beta activity in bifrontal areas *(left map). Right map,* Sudden decrement of beta activity due to an injection of a benzodiazepine antagonist (flumazenil). Beta activity at C4 has been preserved

Pathological EEG and/or EP changes may be diminished or even eliminated in patients after drug administration. If this is accompanied by an improvement in psychopathology or mental performance, the medication does not only show an effect upon CNS function but may also be specifically efficacious. An example of drug efficacy was shown in Fig. 47 in a patient suffering from a progressive paralysis. In the untreated state substantial slow activity occurred in frontocentral areas. Neuropsychological testing revealed a moderately severe organic brain syndrome. After administration of penicillin over a long enough period the slow activity disappeared, and psychological test performance improved. Some disease entities display characteristic EEG and EP features in the untreated versus treated state, thus allowing the description of group data. An example was shown in Fig. 63, where a series of patients was treated with an antidepression drug for 4 weeks.

9.7.2 EP Mapping After Administration of Drugs

Mapping has increased the interest of EP in drug evaluation and led to the introduction of exogenous and endogenous components of auditory, visual, and somatosensory EPs in this field (Maurer et al. 1988 d). An example of a change in topography in an acoustically P300 was presented above in Fig. 50, where a patient with Wilson's disease was treated with d-penicillamine.

EEG features in a group of ten patients with MDD were described above (Sect. 9.6.2.3). P300 measurements were carried out at the same time in the untreated state and again 4 weeks after continuous treatment with a daily dosage of 3 × 100 mg moclobemide (Maurer et al. 1988). Before treatment P300 topogra-

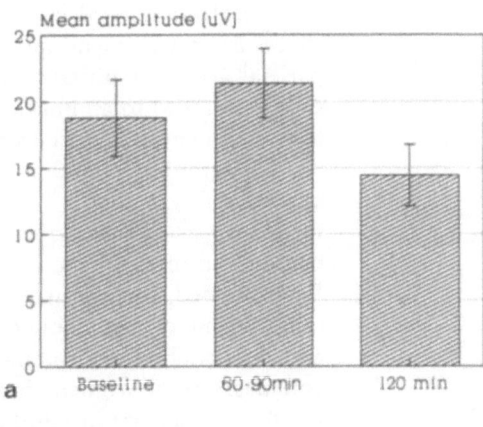

a

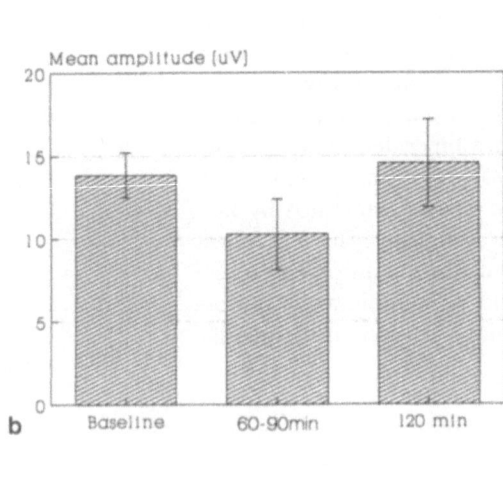

b

I SEM
Mean

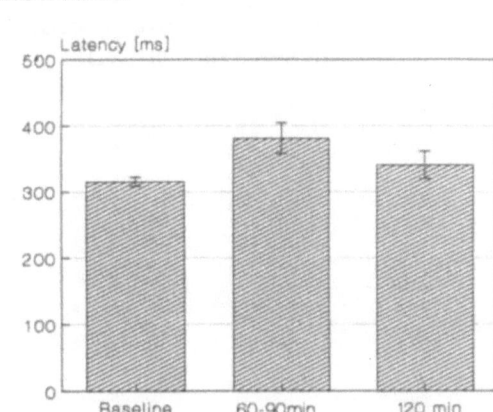

c

Fig. 66 a–d. Bar diagrams indicating P300 amplitudes and
latencies before, after 60–
90 min, and after 120 min of
injection of physostigmine and
biperiden. a Increase of P300
amplitude after drug injection
of physostigmine during a
time period of 60–90 min.
b Decrease of P300 amplitude
after drug injection of biperiden during a time period of
90 min. c Increase in P300 latency after drug injection of
biperiden during a time period
of 90 min. d Results of psychological testing ("hose figures"). Increase of correct
answers after physostigmine
(stripes), corresponding decrease after biperiden (points)

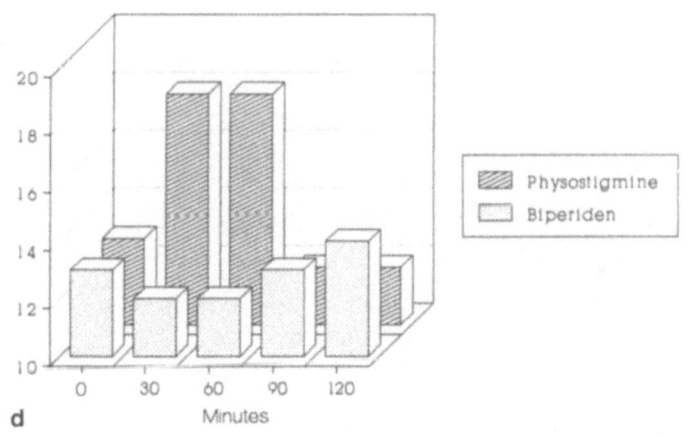

d

phy revealed a normal P300 pattern but slightly decreased amplitudes; after treatment an increase of P300 amplitude in the parietal area was observed (Fig. 63 c, d). The increase in amplitude was linked to an improvement in mood (Hamilton Score before treatment, 27; after treatment, 10).

Topographical P300 studies are valuable in studying central bioavailability not only in cortical but also in allocortical structures such as amygdala, entorhinal area, and hippocampal formation as a result of the assumed allocortical origin of this wave. We examined, for example, P300 behavior after intravenous administration of the cholinergic-acting substance physostigmine and the anticholinergic-acting substance biperiden in six healthy volunteers (two women, four men; 25–31 years of age). As in Fig. 66, an increase in P300 amplitude was accompanied by better performance on neuropsychological testing. The P300 test may be used to determine whether so-called cognition enhancers also develop potency in the limbic and paralimbic areas important for cognitive processing and mainly affected in dementive disorders such as DAT.

10 Advanced Methods

Almost all the methods described in this atlas for signal analysis, data evaluation, and statistical calculations are similar to procedures known in the field of quantitative EEG analysis. We therefore recommend the guidelines and literature concerning on quantitative EEG (Gevins and Remond 1987) and discuss only briefly advanced methods that are closely related to multichannel recordings in combination with mapping.

10.1 Dipole Source Estimation

Electrical fields measured on the scalp are created by electrical currents within the brain. These electrical currents can be described simplistically as dipoles. In dipole source estimation one calculates the location, strength, and direction of one or more equivalent dipoles. For calculation of characteristic sources a number of assumptions must be made in the model about head and sources. The model of the head describes the potential distribution on the scalp according to the underlying intracranial dipoles (forward problem). A two shell (scalp and skull) or a three shell (scalp, skull, and brain) spherical head model is generally used. The source is usually modeled as an equivalent dipole since its fields represent an equivalent description of the compound activity of all neuronal elements in their vicinity that are orientated parallel to the dipole axis. The equivalent dipole can be described either (a) as a moving dipole where the solution describes the field distribution at a particular slice in time or (b) as a so-called spatiotemporal dipole (Scherg and von Cramon 1985), where sources are assumed to remain active during the period of analysis, with changes in their level of activity but not in their location or orientation. The number of sources active during the period analyzed by the spatiotemporal method can be determined by principal component analysis or singular value decomposition. Both methods can be used not only to localize sources but to reduce EEG and EP data in an effective and sensible way.

10.2 Neurometrics

The field of neurometrics has been developed by John and co-workers (John 1989; John et al. 1977, 1987, 1988, 1989) as an aid in differential diagnosis of various brain dysfunctions. This involves statistical analysis of standardized, quantitative electrophysiological features relative to a set of normative data. These features include univariate and multivariate indicators of absolute power, relative power, mean frequency, and coherence and asymmetry between homologous leads. Data are normalized and standardized using z scores. Fisch and Pedley (1989) have argued against this method, however as the sole diagnostic tool for differential diagnosis.

10.3 Determining Differences Between Maps

Lehmann and Skrandies (1980, 1984) proposed the mean standardized "dissimilarity index" between all recording points of the voltages in two maps, based on the average reference as an index for map comparison. The higher the result (standard deviation), the greater is the dissimilarity of maps.

Tatsuno et al. (1988) developed a method for evaluation of differences betwen the maps of two groups using the Mahalanobis distance calculated from polynomial coefficient vectors obtained from unbiased polynomial interpolation. The greater the Mahalanobis distance, the more prominent are the pattern differences.

Desmedt and Chalkin (1989) presented a simple method for calculating the topographic similarity between two maps in the time domain. Values of the estimator z can vary between $+1$ and -1, with $Z = +1$ indicating similar topography and $Z = -1$ a dissimilar inverse topography between two maps.

References

Adachi-Usami E, Lehmann D (1983) Monocular and biocular evoked average potential field topography: upper and lower hemiretinal stimuli. Exp Brain Res 50: 341–346

Adey WR, Dunlop CW, Hendrix CE (1960) Hippocampal slow waves: distribution and phase relations in the course of approach learning. AMA Arch Neurol 3: 74–90

Adrian ED, Matthews BHC (1934) The Berger rhythm potential changes from the occipital lobes in man. Brain 57: 355–385

Adrian ED, Yamagiva K (1935) The origin of the Berger rhythm. Brain 58: 323–351

Ahn S, Jordan CW, Nuwer MR, Marcus DR, Moore WS (1988) Computed electroencephalographic topographic brain mapping. A new and accurate monitor of cerebral circulation and function for patients having carotoid endarterectomy. J Vascular Surg 8: 247–254

Allison T, Matsumiya Y, Goff GD, Goff WR (1977) Topography of human visual evoked potentials. Electroencephal Clin Neurophysiol 42: 185–197

American Electroencephalographic Society (1987) Statement on the clinical use of quantitative EEG. J Clin Neurophysiol 4: 197

Ananiev VM (1956) The electroencephaloscope. Fiziol Zh 42: 981–988

Anderer P, Saletu B, Kinsberger K, Semlitsch H (1987) Topographic brain mapping of EEG in neuropsychopharmacology. I. Methodological aspects. Methods Exp Clin Pharmacol 9: 371–384

Andreasen BC, Olsen S (1982) Negative vs. positive schizophrenia: definition and validation. Arch Gen Psychiatry 39: 789–799

Ashida H, Tatsuno J, Okamoto J, Maru E (1984) Field mapping of EEG by unbiased polynomial interpolation. Comput Biomed Res 17: 267–276

Baldlock GR, Walter WG (1946) A new electronic analyzer. Electron Engin 18: 339–344

Beaubernard C, Minot R, Macher JP (1987) Lithium clinical study by brain electrical-activity-mapping a case report. Pharmacopsychiatry 20: 197–202

Bechterewa NP, Vvedenskaia IV, Dubikaitis YV, Stepanova TS, Ovnatov BS, Usov VV (1963) Localization of focal brain lesions by electroencephalography. Electroencephal Clin Neurophysiol 15: 177–196

Berger H (1929) Über das Elektrenkephalogramm des Menschen. Arch Psychiatr Nervenkr 87: 527–570

Berger H (1933) Über das Elektrenkephalogramm des Menschen. Arch Psychiatr Nervenkr 99: 555–574

Berger MS, Kincaid J, Ojeman GA, Lettich E (1989) Brain mapping techniques to maximize resection, safety and seizure control in children with brain tumors. Neursurgery 25: 786–792

Bertrand I, Lacape RS (1943) Theorie de l'éléctroencéphalogramme: Etats élémentaires. Doin, Paris

Bertrand O, Perrin F, Pernier JA (1985) A theoretical justification of the average reference in topographic evoked potential studies. Electroencephal Clin Neurophysiol 42: 462–464

Bickford RG (1989) EEG color mapping (toposcopy) advantages and pitfalls, clinical and research perspectives. Am J EEG Tech 29: 19–28

Bickford RG, Billinger TW, Fleming NI, Stewart F (1972) The compressed spectral array (CSA). A pictorial EEG. Proc San Diego Biomed Symp 11: 365–370

Bourne JR, Childers DG, Perry NW (1971) Topological characteristics of the visual evoked response in man. Electroencephal Clin Neurophysiol 30: 423–436

Borg E, Spens KE, Tonnquist I (1988) I. Auditory brain map, effects of age. Scand Audiol 30: 161–164

Borg E, Spens KE, Tonnquist I, Rosen S (1988) Brain map. New possibilities in diagnosis of central auditory disorders? Acta Otolaryngol Stockh 103: 612

Bostem F, Degossely M (1978) Spectral analysis of alpha rhythm during Schultz's autogenic training. Contemp Clin Neurophysiol 34 [Suppl]: 181–190

Bourne JR, Childers DG, Perry NW (1971) Topological characteristics of the visual evoked response in man. Electroencephal Clin Neurophysiol 30: 423–436

Brazier MAB, Casby JU (1956) Some applications of correlation analysis to clinical problems in electroencephalography. Electroencephal Clin Neurophysiol 8: 325–331

Breitling D, Guenther W, Rondot P (1987) Auditory perception of music measured by brain electrical mapping. Neurophysiologia 25: 765–774

Breslau J, Starr A, Sicotte N, Higa J, Buchsbaum MS (1989) Topographic EEG changes with normal aging and SDAT. Electroencephal Clin Neurophysiol 72: 281–289

Brown WS, Lehmann D (1979) Verb and noun meaning of homophone words activate different cortical generators: a topographical study of evoked potential fields. Exp Brain Res 2 (Suppl): 159–168

Buchsbaum MS, King AC, Cappelletti J, Coppolla R, van Kammen DP (1982 a) Visual evoked potential topography in patients with schizophrenia and normal controls. Adv Biol Psych 9: 50–56

Buchsbaum MS, Mendelson WB, Duncan WC, Coppola R, Kelsoe J, Gillin JC (1982 b) Topographic cortical mapping of EEG sleep stages during daytime naps in normal subjects. Sleep 5 (3): 248–255

Buchsbaum MS, Rigal F, Coppola R, Cappelletti J, King C, Johnson J (1982 c) A new system for grey level surface distribution maps of electrical activity. Electroencephal Clin Neurophysiol 53: 237–242

Buchsbaum MS, Awsare SV, Holcomp HH, Delisi LE, Hazlett E, Carpenter WT, Pickar D, Morihisa JM (1986) Topographic differences between normals and schizophrenics: the N120 evoked potential component. Neuropsychobiology 15: 1–6

Carl G, Dierks T, Engelhardt W, Maurer K (1989) Topographic EEG features during deep isoflurane anesthesia in patients with major depressive disorder. In: Maurer K (ed) Topographic brain mapping of EEG and evoked potentials. Springer, New York Berlin Heidelberg, pp 259–264

Caton R (1875) The electrical currents of the brain. Br Med J 2: 278

Chatrian GE, Leetich E, Nelson PA (1988) Modified nomenclature for the "10%" electrode system. J Clin Neurophysiol 5: 183–186

Chweitzer A, Geblewicz E, Liberson W (1936) Etude de l'electroencephalogramme humain dans un cas d'intoxication mescalinique. Ann Psychol Paris 37: 94–119

Coburn KL, Moreno MA (1988) Facts and artifacts in brain electrical activity mapping. Brain Topography 1: 37–45

Comacchio F, Grandori F, Magnavita V, Martini A (1988) Topographic brain mapping of middle latency auditory evoked potentials in normal subjects. Scand Audiol 30 (Suppl): 165–172

Cooley WJ, Tukey JW (1965) An algorithm for the machine calculation of complex Fourier series. Math Comput 19: 297–301

Cooper R, Shipton HW, Shipton J, Walter VJ, Walter WG (1957) Spatial and temporal identification of alpha activities in relation to individual mental states, by means of the 22-channel helical scan toposcope. Electroencephal Clin Neurophysiol 9: 375

Coppola R, Buchsbaum MS, Rigal F (1982) Computer generation of surface distribution maps of measures of brain activity. Comput Biol Med 12/3

Deiber MP, Ibanez V, Fischer C, Perrin F, Mauguire F (1988) Sequential mapping favours the hypothesis of distinct generators for Na and Pa middle latency auditory evoked potentials. Electroencephal Clin Neurophysiol 71: 187–197

De Mott DW (1966) Cortical micro-toposcopy. Med Res 5: 23–29

Desmedt JE, Bourguet M (1985) Color imaging of parietal and frontal somatosensory potential fields evoked by stimulation of median or posterior tibial nerve in man. Electroencephal Clin Neurophysiol 62: 1–17

Desmedt JE, Chalklin V (1989) New method for titrating differences in scalp topographic patterns in brain evoked potential mapping. Electroencephal Clin Neurophysiol 74: 359–366

Desmedt JE, Tomber C (1989) Mapping early somatosensory evoked potentials in selective attention: critical evaluation of control conditions used for titrating by difference the cognitive P30, P40, P100 and N140. Electroencephal Clin Neurophysiol 74: 321–346

Desmedt JE, Nguyen TH, Bourguet M (1987) Bit-mapped color imaging of human evoked potentials with reference to the N20, P22, P27 and N30 somatosensory responses. Electroencephal Clin Neurophysiol 68: 1–19

Dierks T, Maurer K, Ihl R (1989 a) Referenzunabhängige Bestimmung von Latenz, Amplitude und Topographie der P300 und 3D-Darstellung bei der Demenz von Alzheimer Typ (DAT). In: Saletu B (ed) Biologische Psychiatrie. Thieme, Stuttgart, pp 207–211·

Dierks T, Maurer K, Ihl R, Schmidtke A (1989) Evaluation and interpretation of topographic EEG data in schizophrenic patients. In: Maurer K (ed) Topographic mapping of EEG and evoked potentials. Springer, New York Berlin Heidelberg, pp 507–517

Dierks T, Maurer K (1989) P300 evoked by an auditory and a visual paradigm and a semantic task. Psychiatr Res 29: 439–441

Dierks T, Maurer K (1990) Reference-free evaluation of auditory evoked potentials P300 in aging and dementia. In: Dostert P et al. (eds) Early markers in Parkinson's and Alzheimer's disease. Springer, Vienna, pp 197–208

Dierks T, Maurer K, Zacher A (1989 b) Brain mapping of EEG in autogenic training (AT). Psychiatr Res 19: 433–434

Dierks T, Perisic I, Froehlich L, Ihl R, Maurer K (1991) Topography of QEEG in dementia of Alzheimer Type (DAT): Relation of severity of disease. Psychiatr Res (in press)

Dietsch G, Berger H (1932) Fourier Analyse von Elektroenkephalogrammen des Menschen. Pflügers Arch 230: 106–112

Drohocki Z (1939) Electrospectrographie des Gehirns. Klin Wochenschr 18: 536–538

Duff TA (1980 b) Topography of scalp recorded potentials by stimulation of the digits. Electroencephal Clin Neurophysiol 49: 452–460

Duff TA (1980 a) Multichannel topographic analysis of human somatosensory evoked potentials. Progr Clin Neurophysiol 7: 69–86

Duffy FH (1981) Brain electrical activity mapping (BEAM): computerized access to complex brain function. Int J Neurosci 13: 55–65

Duffy FH (1986) Topographic mapping of brain electrical activity. Butterworth, Boston

Duffy FH (1989) Clinical value of topographic mapping and quantified neurophysiology. Arch Neurol 46: 1133–1134

Duffy FH, Maurer K (1989) Establishment of guidelines for the use of topographic mapping in clinical neurophyisology: a philosophical approach. In: Maurer K (ed) Topographic mapping of EEG and evoked potentials. Springer, New York Berlin Heidelberg, pp 3–10

Duffy FH, Burchfield JL, Lombroso CT (1979) Brain electrical activity mapping (BEAM): a method for extending the clinical utility of EEG and evoked potential data. Ann Neurol 5: 309–321

Duffy FH, Bartels PH, Burchfield JL (1981) Significance probability mapping: an aid in the topographic analysis of brain electrical mapping. Electroencephal Clin Neurophysiol 51: 455–462

Duffy FH, Albert MS, McAnulty G, Garvey AJ (1984 a) Age related differences in brain electrical activity of healthy subjects. Ann Neurol 16: 430–438

Duffy FH, Albert MS, McAnulty G (1984 b) Brain electrical activity in patients with presenile and senile dementia of the Alzheimer type. Ann Neurol 16: 430–438

Duffy FH, Jensen F, Erba G, Burchfiel JL, Lombroso CT (1984 c) Extraction of clinical information from electroencephalographic background activity: the combined use of brain electrical activity mapping and intravenous sodium penthal. Ann Neurol 15: 22–30

Edwards L, Drasdo N (1987) Scalp distribution of visual evoked potentials to foveal pattern and luminance stimuli. Documenta Ophthalmologica 66: 301–311

Engelhardt W, Carl G, Dierks T, Maurer K (1989) EEG mapping in anesthesia. Basis 2 (3): 2–4

Engelhardt W, Kullmann F, Dierks T, Maurer K (1989) Brain electrical activity mapping: während Narkoseeinleitung mit Thiopental. Anaesthesist 38 (Suppl 1): S141 P4.2

Engelhardt W, Carl G, Dierks T, Maurer K (1990) EEG mapping during Isoflurane anesthesia for treatment of mental depression. J Clin Monitoring 7: 23–29

Eppstein CM, Brickley GP (1985) Interelectrode distance and amplitude of the scalp EEG. Electroencephal Clin Neurophysiol 60: 287–292

Erzigkeit H (1977) Manual zum Syndrom Kurztest. Vless, Munich

Etevenon P, Gaches J (1984) Quantitative EEG maps in neuropsychiatry: problems and perspectives. Clin Neuropharmacol 1: 122–123

Faux SF, Torello MW, McCarley RW, Shenton ME, Duffy FH (1988) P300 in schizophrenia: confirmation and statistical validation of temporal region deficit in P300 topography. Biol Psychiatr 23: 776–790

Findji F, Catani P, Liard C (1981) Topographical distribution of delta rhythms during sleep: evolution with age. Electroencephal Clin Neurophysiol 51: 659–665

Fisch BJ, Pedley TA (1989) The role of quantitative EEG topographic mapping or 'neurometrics' in the diagnosis of psychiatric and neurological disorders: the cons. Electroencephal Clin Neurophysiol 73: 5–9

Freeman WJ, Maurer K (1989 a) Advances in brain theory give new directions of the use of technologies of brain mapping in behavioral studies. In: Maurer K (ed) Topographic brain mapping of EEG and evoked potentials. Springer, New York Berlin Heidelberg, pp 118–126

Freeman WJ, Maurer K (1989 b) Images and imaginings from computerized brains. Psychiatr Res 29: 239–245

Fritze J, Dierks T, Maurer K (1989) EEG-mapping during cholinergic drug challenge by RS-86. In: Maurer K (ed) Topographic brain mapping of EEG and evoked potentials. Springer, New York Berlin Heidelberg, pp 518–521

Fuenfgeld EW (1989) Brain electrical activity mapping in different stages of SDAT. Psychiatr Res 29: 411–412

Garber HJ, Weilburg JB, Duffy FH, Manschreck TC (1989) Clinical use of topographic brain electrical activity mapping in psychiatry. J Clin Psychiatr 50: 205–211

Gevins A (1989) Recent advances in mapping cognition. Symposium on Topographic EEG and EP analysis, Aosta, Italy, September 7–10

Gevins AS, Remond A (eds) (1987) Handbook of electroencephalography and clinical neurophysiology. Vol 1: Methods of analysis of brain electrical and magnetic signals. Elsevier, Amsterdam

Goff GD, Matsumiya Y, Allison T, Goff WR (1977) The scalp topography of human somatosensory and auditory evoked potentials. Electroencephal Clin Neurophysiol 42: 57–76

Goff WR, Rosner BS, Allison T (1962) Distribution of cerebral somatosensory evoked responses in normal man. Electroencephal Clin Neurophysiol 14: 697–713

Grandori F (1986) Field analysis of auditory evoked brainstem potentials. Hear Res 21: 51–58

Grass MA, Gibbs FA (1938) Fourier transform of the EEG. J Neurophysiol (Springfield) 1: 521–526

Guenther W, Breitling D (1985) Predominant sensorimotor area left hemisphere dysfunction in schizophrenia measured by brain electrical activity mapping. Biol Psychiatr 20: 515–532

Guenther W, Breitling W, Banquet JP, Marcie P, Rondot P (1986) EEG mapping of left hemisphere dysfunction during motor performance in schizophrenia. Biol Psychiatr 21: 249–262

Halliday AM, Barrett G, Halliday E, Michael WF (1977) The topography of the pattern evoked potential. In: Desmedt JE (ed) Visual evoked potentials in man: new developments. Clarendon, Oxford

Hamburger H (1989) A battery approach to clinical utilisation of topographic brain mapping. In: Maurer K (ed) Topographic mapping of EEG and evoked potentials. Springer, Berlin Heidelberg New York Toyko, pp 167–184

Harner RN, Ostergren KA (1978) Computed EEG topography. In: Contemporary Clin Neurophysiol (EEG Suppl 34)

Harner RN, Jackel RA, McWhinney-Hee MR, Sussman NM (1987) Computed EEG topography in epilepsy. Rev Neurol Paris 457–461

Hegerl U, Klotz S, Ulrich G (1985) Späte akustisch evozierte Potentiale – Einfluß von Alter, Geschlecht und unterschiedlichen Untersuchungsmethoden. Z EEG-EMG 16: 171

Hermann WM, Kubicki St, Künkel H, Kugler J, Lehmann D, Maurer K, Rappelsberger P, Scheuler W (1989) Empfehlungen der deutschen EEG-Gesellschaft für das Mapping von EEG-Parametern (EEG- and EP-Mapping) Z EEG-EMG 20: 125–132

Hjorth B (1970) EEG-analysis based on time domain properties. Electroencephal Clin Neurophysiol 29: 306–310

Hjorth B (1973) The physicial significance of time domain descriptions in EEG-analysis. Electroencephalogr Clin Neurophysiol 34: 321–325

Hjorth B (1975) An on-line transformation of EEG scalp potentials into orthogonal source derivations. Electroencephal Clin Neurophysiol 39: 526–530

Hjort B (1980) Source derivation simplifies topographical EEG interpretation. Am J EEG Technol 208: 121–132

Hughes JR, Miller JK (1989) An example of the possible clinical usefulness of topographic EEG displays. Clin Electroencephal 20: 39–44

Ihl R, Maurer K, Dierks T, Froelich L, Perisic I (1989) Staging in dementia of the Alzheimer type – topography of electrical brain activity reflects the severity of the disease. Psychiatr Res 29: 399–401

Irrgang U, Höller L (1981) Polygraphic recording in the EEG laboratory. J Electrophys Technol 7: 98–111

Itil TM (1961) Electroenzephalographische Befunde zur Klassifikation neuro- und thymoleptischer Medikamente. Med Exp 5: 347–363

Jacobsen GP, Grayson AS (1988) The normal scalp topography of middle latency auditory evoked potential Pa component following monaural click stimulation. Brain Topography 1: 29–36

Jasper H (1958) Report of committee on methods of clinical exam in EEG. Electroencephal Clin Neurophysiol 10: 370–375

Jerrett SA, Corsak J (1988) Clinical utility of topographic EEG brain mapping. Clin Electroencephal 19: 134–143

John ER (1989) The role of quantitative EEG topographic mapping or 'neurometrics' in the diagnosis of psychiatric and neurological disorders: the pros. Electroencephal Clin Neurophysiol 73: 2–4

John ER, Karmel BZ, Corning WC, Easton P, Brown D, Ahn H, John M, Harmony T, Prichep L, Toro A, Gerson I, Bartlett F, Thatcher F, Kaye H, Valdes P, Schwarz E (1977) Neurometrics: numerical taxonomy identifies different profiles of brain functions within groups of behaviorally similar people. Science 196: 1393–1410

John ER, Prichep LS, Easton P (1987) Normative data banks and neurometrics: basic concepts, current status and clinical applications. In: Remond A, Lopes da Silva F (eds) Computer analysis of EEG and other neurophysiological variables: clinical applications. Elsevier, Amsterdam (EEG handbook, vol 3)

John ER, Prichep LS, Friedman J, Easton P (1988) Neurometrics: computer-assisted differential diagnosis of brain dysfunction. Science 293: 162–169

John ER, Prichep LS, Friedman J, Easton P (1989) Neurometric topographic mapping of EEG and evoked potential features: application to clinical diagnosis and cognitive evaluation. In: K. Maurer (ed) Topographic brain mapping of EEG and evoked potentials. Springer, New York Berlin Heidelberg, pp 90–117

Jones SJ, Power CN (1984) Scalp topography of human somatosensory evoked potentials: the effect of interfering tactile stimulation applied to the hand. Electroencephal Clin Neurophysiol 58: 25–36

Kahn EM, Weiner RD, Brenner RP, Coppola R (1988) Topographic maps of brain electrical activity pitfalls and precautions. Biol Psychiatr 23: 628–636

Kakigi R, Shibasakin H (1983) Scalp topography of the short latency somatosensory evoked potentials following posterior tibiale nerve stimulation in man. Electroencephal Clin Neurophysiol 56: 430–437

Karniski W, Clifford, Blair R (1989) Topographical and temporal stability of the P300. Electroencephal Clin Neurophysiol 72: 373–383

Karson CN, Coppola R, Morihisa JM, Weinberger DR (1987) Computed electroencephalographic activity mapping in schizophrenia. Arch Gen Psychiatr 44: 514–517

Katzelson RD (1981) EEG recording, electrode placement, and aspects of generator localization. In: Nunez PL, Katzelson RD (eds) Electric fields of the brain: the neurophysics of EEG. Oxford University Press, London, pp 176–213

Kitani Y, Watanabe Y, Fujita T (1985) Monitoring of EEG activity during NLA anaesthesia recorded on topographic computerized display map. Electroencephal Clin Neurophysiol 61: 598

Kohrmann MH, Sugioka C, Huttenlocher PR, Spire JP (1989) Intersubject versus intra-subject variance in topographic mapping of the electroencephalogram. Clin Electroencephal 20: 248–253

Künkel H (1972) Simultane Viel-Kanal-On-Line-EEG-Analyse in Echtzeit. Z EEG-EMG 3: 29

Künkel H, Luba A, Niethardt P (1976) Topographic and psychosomatic aspects of spectral EEG analysis of drug effect. In: Kellaway P, Petersen I (eds) Quantitative analytic studies in epilepsy. Raven Press, New York, pp 207–223

Lai CW (1986) The effect of eye/hand dominance on topographic distribution of visual evoked potentials. Electroencephal Clin Neurophysiol 64: 82

Lee S, Buchsbaum MS (1987) Topographic mapping of EEG artifacts. Clinical EEG 38: 3–10

Lechner H, Niederkorn K, Logar C, Schmidt R (1989) Topographic EEG brain mapping in cerebrovascular disease and dementia. Neurologija 38: 3–10

Lehmann D (1971) Multichannel topography of human alpha EEG fields. Electroencephal Clin Neurophysiol 31: 439–449

Lehmann D (1986) Spatial analysis of EEG and evoked potential data. In: Duffy FH (ed) Topographic mapping of brain electrical activity. Butterworths, Boston

Lehmann D (1987) Principles of spatial analysis. In: Gevins A, Remond A (eds) Handbook of electroencephalography and clinical neurophysiology, vol 1: Methods of analysis of brain electrical and magnetic signals. Elsevier, Amsterdam, pp 309–354

Lehmann D (1989) The view of an EEG-EP mapper. Brain Topography 1: 77–78

Lehmann D, Skrandies W (1980) Reference-free identification of components of checkerboard-evoked multichannel potential fields. Electroencephal Clin neurophysiol 48: 609–621

Lehmann D, Skrandies W (1984) Spatial analysis of evoked potentials in man an overview. Progr Neurobiol 23: 227–250

Lemieux JF, Blume FT (1986) Topographical evolution of spike wave complexes. Brain Res 373/1–2

Lemieux JF, Vera RS, Blume WT (1984) Technique to display topographical evolution of EEG events. Electroencephal Clin Neurophysiol 58: 565–568

Lilly CA (1950) A method of recording the moving electrical potential gradients in the brain: the 25 channel bavatron and electroiconograms. Conference on electronics in nucleonics and medicin. Am Inst Electron, pp 37–43

Lilly JC (1954) Instantaneous relations between activities of closely spaced zones on cerebral cortex: Electrical figures during responses of spontaneous activity. Ann J Physiol 176: 493–504

Lowitzsch K, Maurer K, Hopf M (1983) Evozierte Potentiale in der klinischen Diagnostik. Thieme, Stuttgart

Lukas SE, Mendelson JH, Woods BT, Mello NK, Teoh SK (1989) Topographic distribution of EEG alpha activity during ethanol-induced intoxication in women. J Stud Alcohol 50: 176–185

Maurer K (ed) (1989) Topographic brain mapping of EEG and evoked potentials. Springer, New York Berlin Heidelberg

Maurer K, Dierks T (1987 a) Brain-Mappping topographische Darstellung des EEG und der evozierten Potentiale in Psychiatrie und Neurologie. Z EEG EMG 18: 4–12

Maurer K, Dierks T (1987 b) Functional imaging of the brain in psychiatry mapping of EEG and evoked potentials. Neurosurg Rev 10: 275–282

Maurer K, Leitner H, Schäfer E (1982) Akustisch evozierte Potentiale (AEP). Enke, Stuttgart

Maurer K, Dierks T, Ihl R (1988 a) Quantitative P300 data and their topography in dementia. In: Samson-Dollfus (ed) Statistics and topography in quantitative EEG. Elsevier, Amsterdam, pp 243–250

Maurer K, Dierks T, Laux G, Rupprecht R, Ihl R (1988 b) Topographic mapping of EEG and auditory P300 in neuropsychopharmacology. Pharmacopsychiatry 21: 338–342

Maurer K, Ihl R, Kuhn W, Dierks T (1988 c) Brain mapping of EEG and EP during physiological aging and in Parkinsons disease, dementia and depression. In: Pruntzek H, Riederer P (eds) Diagnosis and preventive therapy of Parkinsons disease. Springer, Vienna New York, pp 117–124

Maurer K, Ihl R, Dierks T (1988 d) Topographie der P300 in der neuropsychiatrischen Pharmakotherapie III. Kognitives P300 feld beim organischem Psychosyndrom (M. Wilson), vor und während einer Therapie mit D-Penicillamin. Z EEG EMG 19: 62–64

Maurer K, Lowitsch K, Stöhr R (1988 e) Evozierte Potentiale: AEP-VEP-SEP. Enke, Stuttgart

Maurer K, Dierks T, Rupprecht R (1989 a) Computerized encephalographic topography (CET) during sleep. Psychiatry Res 29: 435–438

Maurer K, Dierks T, Ihl R, Laux G (1989 b) Mapping of evoked potentials in normals and patients with psychiatric diseases. In: Maurer K (ed) Topographic brain mapping of EEG and evoked potentials. Springer, New York Berlin Heidelberg, pp 458–473

Maurer K, Lowitsch K, Stöhr M (1989 c) Evoked potentials. Decker, Torronto

Maurer K, Riederer P, Heinsen H, Beckmann H (1989 d) Altered P300 topography due to functional and structural disturbances in the limbic system in dementia and psychosis. Psychiatr Res 29: 391–393

McKhann G, Drachman D, Folstein M, Katzman R, Price D, Stadlan EM (1984) Clinical diagnosis of Alzheimer's disease: report of the NINCDS-ADRDA work-group under the auspices of department of health and human services task force on Alzheimer's disease. Neurology 34: 939–944

Mezan I, Lesevre N, Remond A (1968) Etude chrono-topographique de la reponse somesthesique evoquee moyenne recueillie sur le scalp par stimulation electrique du nerf median. Rev Neurol 119 (3): 288–295

Morihisa JM, McAnulty GB (1985) Structure and function: brain electrical activity mapping and computed tomography in schizophrenia. Biol Psychiatr 20/1

Morihisa JM, Duffy FH, Wyatt RJ (1983) Brain electrical activity mapping (BEAM) in schizophrenic patients. Arch Gen Psychiatry 40: 719–726

Morstyn R, Duffy FH, McCarley RW (1983 a) Altered P300 topography in schizophrenia. Arch Gen Pschiatr 40: 729–734

Morstyn R, Duffy FH, McCarley RW (1983 b) Altered topography of EEG spectral content in schizophrenia. Electroencephal Clin Neurophysiol 56: 263–271

Motokava K (1944) Die Verteilung der elektrischen Aktivität auf der Kopfschwarte und ihre Beziehung zur Cytoarchitektonik der Großhirnrinde des Menschen. Jpn J Med Sci Biol 3 (10): 99–111

Nagata K (1989) Topographic EEG in brain ischemia correlation with blood flow and metabolism. Brain Topography 1: 97–106

Nagata K, Yunoki K, Araki G, Mizukami M (1984) Topographic electroencephalographic study of transient ischemic attacks. Electroencephal Clin Neurophysiol 58: 291–301

Nagata K, Gross CE, Kindt GW, Geier JM, Adey GR (1985) Topographic electroencenphalographic study with power ratio index mapping in patients with malignant brain tumors. Neurosurgery 17: 613–619

Nunez PL (1988) Methods of estimate spatial properties of dynamic cortical source activity. In: Pfurtscheller G, Lopes da Silva F. H. (Eds.) Functional Brain Imaging. Huber, Toronto, pp 3–10

Nuwer MR (1985) A comparison of the analyses of EEG and evoked potentials using colored bars in place of colored heads. Electroenceph Clin Neurophysiol 61: 310–313

Nuwer MR (1987) Recording electrode site nomenclature. J Clin Neurophysiol 4: 121–133

Nuwer MR (1988 a) Frequency analysis and topographic mapping of EEG and evoked potentials in epilepsy. Electroencephal Clin Neurophysiol 69: 118–126

Nuwer MR (1988 b) Quantitative EEG: I. Techniques and problems of frequency analysis and topographic mapping. J Clin Neurophysiol 5: 1–43

Nuwer MR (1988 c) Quantitative EEG: II. Frequency analysis and topographic mapping in clinical settings. J Clin Neurophysiol 5: 45–85

Nuwer MR (1989) Uses and abuses of brain mapping. Arch Neurol 46: 1134–1136

Nuwer MR, Jordan SE (1987) The centrifugal effect and other spatial artifacts of topographic mapping. J Clin Neurophysiol 4: 321–326

Nuwer MR, Jordan SE, Ahn SS (1987) Evaluation of stroke using EEG frequency analysis and topographic mapping. Neurology 37: 1153–1159

Offner FF (1950) The EEG as potential mapping: the value of the average monopolar reference. Electroencephal Clin Neurophysiol 2: 215–216

Ormejohnson DW, Gelderloos P (1988) Topographic EEG brain mapping during yogic flying. Int J Neurosci 38: 427–434

Perrin F, Pernier J, Bertrand O, Giard MH, Echallier JF (1987) Mapping of scalp potentials by surface spline interpolation. Electroencephal Clin Neurophysiol 66: 75–81

Petsche H (1952) Das Vektor-EEG, ein neuer Weg zur Klärung hirnelektrischer Vorgänge. Z Nervenheilk Wien 5: 304–320

Petsche H (1962) Pathophysiologie und Klinik des Petit Mal. Toposkopische Untersuchungen zur Phänomenologie des Spike Wave Musters. Z Nervenheilk Wien 19 (4): 345–442

Petsche H (1976) Topography of the EEG: survey and prospects. Clin Neurol Neurosurg 79: 15–28

Petsche H, Marko A (1955) Toposkopische Untersuchungen zur Ausbreitung des Alpharhythmus. Z Nervenheilk Wien 12: 87–100

Petsche H, Stumpf SL (1960) Topographic and toposcopic study of origin and spread of the regular synchronized arousal pattern in the rabbit. Electroencephal Clin Neurophysiol 589–600

Pfefferbaum A, Ford JM, White PM, Roth WT (1989) P3 in schizophrenia is affected by stimulus modality, response requirements, medication status, and negative symptoms. Arch Gen Psychiatry 46: 1035–1044

Pfurtscheller G (1986) Event-related desynchronization mapping. Visualization of cortical activation patterns. In: Duffy PH (ed) Topographic mapping of brain electrical activity. Butterworths, Boston, pp 95–111

Pfurtscheller G, Lopes da Silva FH (1988) Functional brain imaging. Huber, Toronto

Pichlmayr I, Lips U, Künkel H (1984) The electroencephalogram in anaesthesia. Springer, Berlin Heidelberg Tokyo, p 29

Pockberger H, Rappelsberger P, Petsche H, Thau K, Kufferle B (1984) Computer assisted EEG topography as a tool in the evaluation of actions of psychoactive drugs in patients. Neuropsychobiology 12: 183–187

Pockberger H, Petsche H, Rappelsberger P, Zidek B, Zapotoczky HG (1985) On-going EEG in depression: a topographic spectral analytical pilot study. Electroencephal Clin Neurophysiol 61: 349–358

Poimann H, Maurer K, Dierks T (1989) Mapping of EEG in patients with intracranial structure lesions. In: Maurer K (ed) Topographic brain mapping of EEG and evoked potentials. Springer, New York Berlin Heidelberg, pp 278–284

Ragot RA, Remond A (1978) EEG field mapping. Electroencephal Clin Neurophysiol 45: 417–421

Rechtschaffen and Kales (1968) A manual of standardized terminology, techniques and storing system for sleep stages of human subjects. Publ Health Service, US Goverment, Washington DC

Reisberg B, London E, Ferris SH, Borenstein J, Scheier L, de Leon MJ (1983) The Brief Cognitive Rating Scale: language, motoric and mood, concommitants in primary degenerative dementia (PDD). Psychopharmacol Bull 19: 702–708

Remond A, Offner F (1952 a) Etudes topographiques de l'activite EEG de la region occipitale. Rev Neurol (Paris) 87: 182–189

Remond A, Offner F (1952 b) A new method for EEG display. Electroencephal Clin Neurophysiol 7: 453–460

Remond A, Delarue R (1959) Le systeme EHP58: electrodes, harnais, placement a 58 electrodes d'enregistrement electroencephalographique. Rev Neurol 101 (3): 261–262

Rodin EA (1988) Computer assisted clinical neurophysiology the role of the technologist. J Electrophysiol Technol 14: 91–108

Rodin E, Cornellier D (1989) Source derivation recordings of generalized spike wave complexes. Electroencephal Clin Neurophysiol

Saletu B, Gruenberger J (1988) Drug profiling by computed electroencephalography and brain maps, with special consideration of Sertraline and its psychometric effects. J Clin Psychiatr 49: 59–71

Saletu B, Anderer P, Kinsberger K, Gruenberger J (1987) Topographic brain mapping of EEG in neuropsychopharmacology. II. Clinical applications (pharmaco EEG imaging). Method Find Exp Clin Pharmacol 9: 385–408

Saletu B, Anderer P, Gruenberger J (1989) EEG brain mapping in gerontopsychopharmacology – on protective properties of pyritinol against hypoxic hypoxidoses. Psychiatr Res 29: 387–390

Saltzberg B, Burch NR (1957) A new approach to signal analysis in electroencephalography. IRE Trans Biomed Eng 8: 24–30

Samson-Dollfus D, Guieu JD, Gotman J, Etevenon P (1989) Statistics and topography in quantitative EEG. Proceedings of the International Workshop on Statistics and Topographics Problems in Quantitative EEG, Rouene France March 6–9 (1988)

Scherg M, von Cramon D (1985) Two bilateral sources of the late AEP as identified by a spatio-temporal dipole model. Electroencephal Clin Neurophysiol 62: 32–44

Shenton ME, Faux SF, McCarley RW, Ballinger R, Coleman M, Duffy FH (1989) Correlations between abnorma lauditory P300 topography and positive symptoms in schizophrenia a preliminary report. Biological Psychiatry 25: 710–716

Skrandies W, Lehmann D (1982) Spatial principal components of multichannel maps evoked by lateral visual halffield stimuli. Electroencephal Clin Neurophysiol 54: 662–667

Spitzer AR, Cohen LG, Fabrikant J, Hallet M (1989) A method for determining optimal interelctrode spacing for cerebral topographic mapping. Electroencephal Clin Neurophysiol 72: 355–361

Taira T, Amano K, Kawamura H et al. (1986) Significance probability mapping of brain electrical activity. Its problem and specified Z-statistic mapping. Neurol Surg 14/3

Tatsuno J, Ashida H, Takao A (1988) Objective evaluation of differences in patterns of EEG topographical maps by Mahalanobis distance. Electroencephal Clin Neurophysiol 69: 287–290

Thau K, Rappelsberger P, Lovrek A, Petsche H, Simhandl C, Topitz A (1989) Effect of lithium on the EEG of healthy males and females. A probability mapping study. Neuropsychobiology 20: 158–163

Thickbroom GW, Davies HD, Carroll WM, Mastaglia FL (1986) Averaging, spatio-temporal mapping and dipole modelling of focal epileptic spikes. Electroencephal Clin Neurophysiol 64/3

Tsuji S, Murai Y (1986) Scalp topography and distribution of cortical somatosensory evoked potentials to median nerve stimulation. Electroencephal Clin Neurophysiol 65: 429–439

Ueno S, Matsuoka S, Mizoguchi T, Nagashima M, Cheng CL (1975) Topographic computer display of abnormal EEG activities in patients with CNS diseases. Memoirs Fac Eng, Kyushu Univ 24 (3): 196–209

Van Toller S, Reed MK (1989) Brain electrical activity topographical maps produced in response to olfactory and chemosensory stimulation. Psychiatr Res 29 (3): 429–430

Walter WG, Shipton HW (1951) A new toposcopic display system. Electroencephal Clin Neurophysiol 3: 281–292

Walter DO (1963) Spectral analysis for electroencephalograms. Mathematical determination of neurophysiological relationship from records of limited duration. Exp Neurol 8: 155–181

Walter DO, Etevenon P, Pidoux B, Tortrat D, Guillou S (1984) Computerized topo-EEG spectral maps: difficulties and perspectives. Neuropsychobiology 11: 264–272

Wong PKH, Gregory D (1988) Dipole fields in rolandic discharges. Am J EEG Technol 28: 243–249

Yamada T, Graff-Radford NR, Kimura J, Dickens QS, Adams HP (1985) Topographic analysis of somatosensory evoked potentials in patients with well localized thalamic infarctions. J Neurol Sci 68: 31–46

Zappoli R, Versari A, Paganini M, et al. (1988) Age differences in contingent negative variation activity of healthy young adults and presenile subjects. Ital J Neurol Sci 9: 219–230

Zee Zang Zao, Gelbin, Remond A (1952) Le champ électrique de oeil. Sem Hop, Paris 28/36: 1–8

Subject Index

activated (dynamic) EEG 50
– –, first movement 50
– –, music 50
– –, words 50
aging 62, 63
–, EEG-topography 62
–, P300-topography 62
alertness 9
aliasing 25
allocortex 88
– amygdala 88
– entorhinal area 88
– hippocampal formation 88
alpha focus 3
alpha rhythm 41
– phase reversal 41
amplitude spectral analysis 28
analog to digital conversion
  (ADC) 23
analysis 40
–, confirmatory 40
–, descriptive 40
–, exploratory 40
analysis of variance (ANOVA) 40,
  54
anesthesia 83
artifacts 18
–, cardiac 20
–, eye blinks 19
–, – movements 19
–, muscles 20
–, perspiration 20
–, pulse-wave 20
–, technical 20
astrocytoma 57
auditory evoked potentials 46
– – –, early 49
– – –, endogenous components
  (P300) 49
– – –, exogenous components
  (N1, P1, N2, P2) 49

background activity 41
bar-format 31
baseline 18
betamax 37
betatron 3

bioequipotency 85
biperiden 88
bits 23
"blink holidays" 21
Bonferroni correction 39
brain electrical activity mapping
  (BEAM) 5, 8
– electromagnetic topography
  (BET) 8
– mapping 1, 8
– tumors 56, 57
brief cognitive rating scale (BRCS)
  9, 64

calibration 11
– bar 35
cerebral bioavailability 85
– blood flow (CBF) 1
cerebrovascular diseases 61
checkerboard 46
cognitive enhancers 85
– performance test 64
common average reference (CAR)
  14
component strength 36
computed tomography (CT) 1
computerized
  electroencephalographic
  topography (CET) 1, 8
contingent negative variation
  (CNV) 49
contour maps 34

d-penicillamine 69, 86
data reduction 39, 40
dementia 54
dementia of Alzheimer type (DAT)
  64, 70
– – –, EEG topography 64
– – –, P300 topography 65, 66
depression 80
–, EEG-mapping 80
–, P300-mapping 80, 81
dipole 89
– source estimation 41, 89
– spatiotemporal 89
discriminant analysis 39

dissimilarity index 90
dose-efficacy relation 85
drowsiness 21
drugs 83
–, EEG-mapping 83
–, EP-mapping 86
DSM III 9
dwell time 23

EEG, activated 50
–, conventional 7
–, microfilm 37
– paper 37
– topography 8
electrodes 11–13
ependymoma 57
epileptic seizures 57, 62
– focus 57
equivalent dipole 40, 89
evoked potentials 45–49
– –, auditory 45–46
– –, chemosensory stimulation 49
– –, drugs 86
– – mapping 35, 35
– – olfactory 49
– – somatosensory 49
– – topography
– –, visual 45–46
eye movements 12
– –, horizontal 12
– –, vertical 12

"far-field potentials" 36, 49
fast Fourier transformation (FFT) 27
frequency domain 41

Gaussian distribution 38
Gaussianity 38
glioblastome 57
global field power (GFP) 36, 40
Goldman reference 55

Hamilton score 76
handedness 9
hard-disk-system 37
head model 89
– –, spherical 89
– –, three shell 89
– –, two shell 89
hemangioma 57
hose figures 87

ICD 9
imaging techniques 1
inflammatory disorders 56
interpolation 9, 31
–, linear 31, 55
–, surface-spline 31

intracranial pressure 56
ischemic insult 60

K-complex 31

Laplacian transformation 17
linear regression 39
lithium 85
luetic infection 67, 69

magnetic disk 37
magnetic resonance imaging (MRI) 1
magnetoencephalography (MEG)
Mahalanobis distance 90
major depressive disorder 67, 70, 76
map features 34
– –, colored 34
– –, contour line 34
– –, gray-scale 34
– –, three-dimensional 34
mapping 1, 7, 8
– amplitude 25
–, EEG 8
–, EP 8
– frequency 26
meningeoma 57
moclobemide 86
multi-infarct dementia (MID) 67, 72
multivariate analysis of variance (MANOVA) 39, 54
muscle movements 12
– –, occipital 12
– –, temporal 12

neuroimaging 1
neuroleptics 85
neurometrics 39, 90
Nyquist frequency 25, 44

optical disk 37

P300, acoustically evoked 49
–, visual evoked 46
Parkinson's disease 67, 70
physostigmine 88
Pick's disease 67, 70
pixels 31
power-spectral analysis 28, 44
principal component analysis (PCA) 40
progressive paralysis 67, 69
pseudoimaging 7
psychoses 54, 72
pyritinol 83

RDC 9
recordings 55

–, bipolar 55
–, longitudinal array 55
–, transversal array 55
–, unipolar array 55
references 14–17
–, common average reference
  (CAR) 14
–, ear 14
–, local average reference 14
–, mastoid 14
–, reference-free techniques 14,
  35
–, single electrode 14

sampling rate 23
scale for assessment of negative
  symptoms (SANS) 9
schizoaffectives disorder 72
schizophrenic disorders 74
–, EEG-mapping 75–79
–, P300-mapping 80
sharp waves 41
significance probability mapping
  (SPM) 39
single photon emission computed
  tomography (SPECT) 1
sleep features 50
slow wave activity 41
somatosensory evoked potential
  49

source derivation (estimation) 17,
  40
spatial domain 9, 31, 41
spike evaluation 31, 41
streamer 37
stroke 61

t-test 40, 54
thiopentone 83
thymoleptics 85
time domain 9, 41
time-efficacy relation 85
toposcope 1, 5
tranquillizer 85
transient ischemic attacks (TIA) 61
transients 62
traumata 56

vascular processes 56
videocassettes 37
visual evoked potential 45
–, flash 45
–, pattern shift 45
visual source localization 55

Wilcoxon test 39, 40
Wilson's disease 67, 69, 86
Winchester drives 37

Z-statistics 31, 38

BRAIN STAR

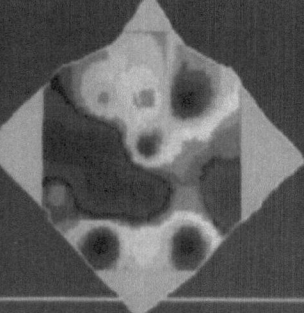

THE NEUROPHYSIOLOGICAL COMPUTER SYSTEM FOR:

Paperless EEG-record and reproduce   Different EEG analysis

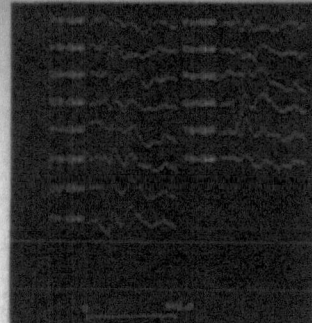

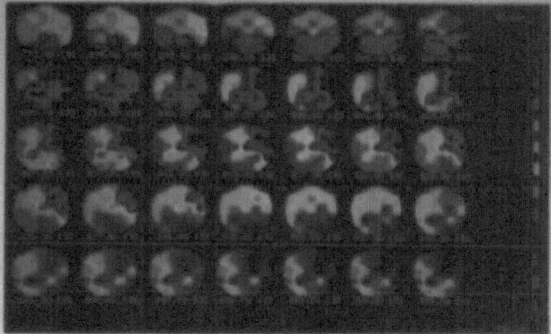

Evoked Potentials   Brain Mapping

Schwind MEDIZINTECHNIK
BRAIN STAR–DIVISION
Barth-Blendinger-Str. 4
D-8520 Erlangen
Telefon: 0 91 31/99 13 44
Telefax: 0 91 31/99 13 74

Additional software available for:
spike-wave analysis "SPIKE-STAR"
sleep analysis "SLEEP-STAR"
EEG-DC amplifier recording
myography and neurography
doppler sonography